THE EFFECTIVE TEACHER SERIES

General editor: Elizabeth Perrott

For series list see pages vii and viii

THE EFFECTIVE TEACHER SERIES

THE EFFECTIVE TEACHING OF BIOLOGY

CHRIS R. BROWN

LONGMAN
LONDON AND NEW YORK

Longman Group Limited,
Longman House, Burnt Mill,
Harlow, Essex CM20 2JE, England
and Associated Companies throughout the world.

Published in the United States of America
by Longman Publishing, New York

First published 1995

ISBN 0 582 095050 PPR

British Library Cataloguing-in-Publication Data

A catalogue record for this book is
available from the British Library

Library of Congress Cataloging-in-Publication Data

Brown, Chris R. (Christopher Rossant), 1940–
The effective teaching of biology / Chris R. Brown.
p. cm. -- (The Effective teacher series)
Includes bibliographical references and index.
ISBN 0-582-09505-0 (paper)
1. Biology--Study and teaching. I. Title. II. Series.
QH315.B823 1995
574'.07'1--dc20 94-33256
CIP

Set by 7E in 10/12 pt Times

Produced through Longman Malaysia, PP

CONTENTS

EDITOR'S PREFACE

This well-established series was inspired by my book on the practice of teaching (*Effective Teaching: a Practical Guide to Improving your Teaching*, Longman, 1982), written for trainee teachers wishing to improve their teaching skills as well as for in-service teachers, especially those engaged in the supervision of trainees. The books in this series have been written with the same readership in mind. However, busy classroom teachers will find that these books also serve their needs as changes in the nature and pattern of education make the in-service training of experienced teachers more essential than in the past.

The rationale behind the series is that professional courses for teachers require the coverage of a wide variety of subjects in a relatively short time so the aim of the series is the production of 'easy to read', practical guides to provide the necessary subject background, supported by references to guide and encourage further reading, together with questions and/or exercises devised to assist application and evaluation.

As specialists in their selected fields, the authors have been chosen for their ability to relate their subjects to the needs of teachers and to stimulate discussion of contemporary issues in education.

The series aims to cover subjects ranging from the theory of education to the teaching of mathematics and from primary school teaching and educational psychology to effective teaching with information technology. It looks at aspects of education as diverse as education and cultural diversity and pupil welfare and counselling. Although some subjects such as the legal context of teaching and the teaching of history are specific to England and Wales, the majority of the subjects such as assessment in education, the effective teaching of statistics and comparative education are international in scope.

Elizabeth Perrott

THE EFFECTIVE TEACHER SERIES

General editor: Elizabeth Perrott

EDUCATION AND CULTURAL DIVERSITY by Edward Hulmes

EFFECTIVE TEACHING; A PRACTICAL GUIDE TO IMPROVING YOUR TEACHING by Elizabeth Perrott

THE EFFECTIVE TEACHING OF ENGLISH by Robert Protherough, Judith Atkinson and John Fawcett

THE EFFECTIVE TEACHING OF MODERN LANGUAGES by Colin Wringe

THEORY OF EDUCATION by Margaret Sutherland

PUPIL WELFARE AND COUNSELLING: AN APPROACH TO PERSONAL AND SOCIAL EDUCATION ACROSS THE CURRICULUM by David Galloway

THE EDUCATION SYSTEM IN ENGLAND AND WALES by John Dunford and Paul Sharp

PRIMARY SCHOOL TEACHING AND EDUCATIONAL PSYCHOLOGY by David Galloway and Anne Edwards

SECONDARY SCHOOL TEACHING AND EDUCATIONAL PSYCHOLOGY by David Galloway and Anne Edwards

CONTEMPORARY DEBATES IN EDUCATION: AN HISTORICAL PERSPECTIVE by Ron Brooks

THE LEGAL CONTEXT OF TEACHING by Neville Harris with Penelope Pearce and Susan Johnstone

TEACHING STATISTICAL CONCEPTS by Anne Hawkins, Flavia Jolliffe and Leslie Glickman

THE EFFECTIVE TEACHING OF MATHEMATICS by Malcolm Simmons

THE TEACHING OF DRAMA IN THE PRIMARY SCHOOL by Brian Woolland

THE EFFECTIVE TEACHING OF RELIGIOUS EDUCATION by Brenda Watson

TEACHING AS COMMUNICATION by Bob Hodge

THE ASSESSMENT OF SPECIAL EDUCATIONAL NEEDS: WHOSE PROBLEM? by David Galloway, Derrick Armstrong and Sally Tomlinson

THE EFFECTIVE TEACHING OF SECONDARY SCIENCE *by John Parkinson*

THE EFFECTIVE TEACHING OF PHYSICAL EDUCATION by Mick Mawer

THE EFFECTIVE TEACHING OF BIOLOGY by Chris R. Brown

AUTHOR'S ACKNOWLEDGEMENTS

Over a long teaching career one encounters many people who help and influence the development of one's ideas. Most significant are the students and I am most grateful to all those with whom I have taught and learned biology. I have drawn on examples of the work of former school, degree and post-graduate students, for which I thank them all.

I am also grateful for some of the ideas which are included in the book to friends and former colleagues in South Wales and to my colleagues in the Science Education Centre at the University of Hull. The principal and staff of the Biology Department of Wyke College, Hull kindly gave permission to include the details which appear in Chapter 4. Other sources are acknowledged in the text. Finally I would like to thank Gill Fincham for typing most of the book and for tolerating, with good grace, my erratic methods of working.

PUBLISHER'S ACKNOWLEDGEMENTS

We are grateful to the following for permission to reproduce copyright material:

Carfax Publishing Company for an extract from an article by P. Murphy in *Cambridge Journal of Education* 1991 21 (2); Institute of Biology, London, for extracts from articles by A Lucas in *Journal of Biological Education* 1987, 2 (1) and B. Longden in *Journal of Biological Education* 1982, 16 (2); Macmillan Magazines Ltd for an extract from an article by J D Watson & F H C Crick in *Nature,* p.737. Copyright 1953 Macmillan Magazines Ltd; John Wiley & Sons Inc for an extract from an article by D F Treagust in *Journal of Research in Science Teaching*, 28 (4) Copyright © 1991, and an extract from an article by Tamir, Choppin & Nussinoritz in *Journal of Research in Science Teaching*, Copyright © 1981. The University of Hull for the Planning Sheet, being our Figure 3.2; The Institute of Biology for the two line drawings in our Figure 6.2, coming from the Autumn 1984 copy of the *Journal of Biological Education* Volume 18, No 3, pp 225 and 224; Don Mackean for Figure 10 on page 36 of *GCSE Biology* (1986) published by John Murray (Publishers) Ltd (our Figure 7.2); Heinemann Publishers (Oxford) Limited and the authors Michael Shayer and Philip Adey for figure 2 (our Figure 7.4) and our Tables 7.2 and 7.3 all from *Towards a Science of Science Teaching: Cognitive Development and Curriculum Demand* © M Shayer and P Adey 1981; Heinemann Publishers (Oxford) Limited for our Figure 8.1 from *Techniques for the Assessment of Practical Skills In Foundation Science* (1983) by T G K Bryce; Nuffield Chelsea Curriculum Trust for our Figure 7.5 from Nuffield Biology *Text II Life and living processings* Longman/Penguin Books 1966, and our Figures 3.1, 4.1, 6.1 and 7.3 from Revised Nuffield Biology *Text 2 Living things in action* Longman 1975; School Curriculum and Assessment Authority for our Figures 8.2 and 8.7 from *SEAC (1992) KS3 School Assessment Folder (Part 2)* and our Figure 8.9 from *SEAC (1993) School Assessment Folder Assessing Science 1;* Open University Press and the author Clive Sutton for the 'fruit' figure on page 61 of *Words, Science and Learning* (1982) (our Figure 8.3); City of Wakefield Metropolitan

District Council for the 'Scientific Investigations' cartoon in Chapter 8; Southern Examining Group for the 'Grid to Show the Relationship Between Strands and Statements of Attainment' (our Figure 8.10) from page 24 of the *GCSE National Curriculum Syllabus – Science for 1994 Examination*; Longman Group UK Limited for our Table 4.3 from *Reading for Learning in the Sciences* (1984) by F. Davies and T. Greene, published by Oliver and Boyd; HMSO for our Table 8.4 from *Science at Age 11: A Review of APU Findings*.

LIST OF ABBREVIATIONS

AIDS	Acquired immune deficiency syndrome
APU	Assessment of Performance Unit
ASE	Association for Science Education
AT	Attainment Target
BSE	Bovine spongiform encephalopathy
CAL	Computer assisted learning
COSHH	Control of Substances Hazardous to Health
DART	Directed activity related to text
DfE	Department for Education
GCSE	General Certificate of Secondary Education
GNVQ	General National Vocational Qualification
HIV	Human immuno-deficiency virus
IOB	Institute of Biology
IT	Information technology
ITT	Initial teacher training
IVF	In-vitro fertilization
KS	Key stage
LEA	Local Education Authority
NCAT	National Curriculum Attainment Target
NCC	National Curriculum Council
NEAB	Northern Examinations and Assessment Board
OFSTED	Office for Standards in Education
PGCE	Post Graduate Certificate in Education
PSE	Personal and social education
QTS	Qualified teacher status
SCAA	School Curriculum and Assessment Authority
SEAC	School Examinations and Assessment Council
SoA	Statement of Attainment
SSCR	Secondary Science Curriculum Review

CHAPTER 1

Introduction

What is effective teaching?

In a book with the title that this one bears, it is logical to start by defining what effective teaching is so that advice about how to achieve it can follow. However an unequivocal answer cannot be given to this question without taking some detours in order to raise other questions on the way. For example is 'effective' the same as 'competent'? Is it possible to distinguish different degrees of competence or effectiveness? And since there are obviously large numbers of different facets attached to the processes of teaching and learning (planning, organizing, managing, controlling, assessing etc.) is an effective teacher equally effective at each facet? And on all occasions? Clearly we might also debate whether your aim in entering teaching is to become effective or do you want to become outstanding, perhaps an exemplary teacher?

My use of the word 'competent' in the previous passage was in order to be able to remind you, that in England and Wales, the Government requires that:

> All newly qualified teachers entering maintained schools should have achieved the knowledge and standards of professional competence necessary to maintain and improve standards in schools.[1]

The expectation that seems to be implied in this Department for Education (DfE) statement is that new entrants to the profession will already have attained standards to match or even better those shown in the schools to which they gain first appointment! Obviously this will not be so; effectiveness builds up with experience. The rate at which this happens differs from person to person. It is determined by many different things including the demands of the context, the extent of the support made available at crucial periods and so on. It may well take several years to become effective, indeed professional development occurs throughout one's professional life.

In the DfE's *Circular No 9/92* on *Initial Teacher Training (Secondary Phase)*, the competences required to be shown by newly qualified teachers are listed under five headings: Subject Knowledge,

Subject Application, Class Management, Assessment and Recording of Pupils' Progress and Further Professional Development. The list is too long to quote in full, it comprises twenty-seven statements or groups of statements. The statements relating to two of these categories are shown as examples.

1. *Subject application.* Newly qualified teachers should be able to:
(a) produce coherent lesson plans that take account of NCATs and of the school's curriculum policies;
(b) ensure continuity and progression within and between classes and in subjects;
(c) set appropriately demanding expectations for pupils;
(d) employ a range of teaching strategies appropriate to the age, ability and attainment level of pupils;
(e) present subject content in clear language and in a stimulating manner;
(f) contribute to the development of pupils' language and communication skills;
(g) demonstrate ability to select and use appropriate resources, including information technology (IT).

2. *Class management.* Newly qualified teachers should be able to:
(a) decide when teaching the whole class, groups, pairs, or individuals is appropriate for particular learning purposes;
(b) create and maintain a purposeful and orderly environment for the pupils;
(c) devise and use appropriate rewards and sanctions to maintain an effective learning environment;
(d) maintain pupils' interest and motivation.

Evaluation of students' mastery of these competences is required in each course of Initial Teacher Training (ITT) by school mentors and Higher Education Institution (HEI) tutors. Qualified Teacher Status (QTS) is awarded to students who show the 'ability to teach effectively and to secure effective learning', and 'the ability to manage pupil behaviour' (*Circular 9/92*, para. 3.4.1). ITT provision in England and Wales is monitored by the Office for Standards in Education (OFSTED), which also undertakes inspection of schools. Exactly the same aspects of teaching competence form the basis for inspection of experienced teachers' classroom work as those on which analysis of trainees is based. Hence a 'satisfactory' level of competence is described in words which are identical to those in *Circular 9/92*. However, OFSTED also describe a level of competence which, going beyond 'satisfactory', constitutes 'very good'. For example, students who are very good in subject application are expected to show additional qualities such as:

imagination and skills in meeting the needs of pupils of different ability levels within the same teaching group; an ability to generate a sense of excitement and enquiry within the subject.[2]

One approach then to defining effective teaching is to focus on the skills or competences that seem to be necessary to do the job. Another possible approach is to turn the issue on its head by identifying what it is that teachers who are widely regarded as being effective actually do in their classrooms. David Treagust, an Australian science educator, made an in-depth study[3] of the teaching of two biology teachers over a period of many weeks. The teachers concerned had been identified as 'outstanding' by their colleagues, Ministry of Education officials and tertiary institution staff. Some selected quotations from his findings are as follows:

One facet common to both teachers was their ability to be always thoroughly prepared and organised . . . students were always aware of exactly what was required of them during each lesson . . . Good management and organisation with an easy flow between segments of the lessons were key features . . . In both teachers' classes, students were all actively engaged in activities for the whole lesson. Students who completed a task in advance of the rest of the class were given extension work. In essence biology classes were busy occasions for students and teachers alike with little opportunity to engage in off-task behaviour . . . all students were encouraged to become involved in class discussion. Motivation took the form of manipulating questioning and the social environment to encourage students to participate successfully, giving effective praise to the class or individuals and assisting students to use time effectively. Interaction was frequent with a strong emphasis being placed on comprehension of concepts. During laboratory activities students were given a considerable degree of independence and they worked at their own pace, usually in pairs. Both . . . had a thorough and comprehensive knowledge of the content they were to teach and had a range of teaching strategies that could be used without a good deal of thought. Their expectations for student performance were high, consistent and firm. Both teachers provided regular feedback to students on their progress and understanding of subject matter.

Clearly there is a high degree of consensus between the competences of the DfE listed earlier and the qualities demonstrated in their classrooms by these two outstanding biology teachers. Is this consensus shared by pupils? In another study[4] questionnaires were given to two classes of each of seven biology trainees who were nearing the end of their teaching practice. Pupils were asked to say how well they thought their teachers carried out different aspects of teaching including: communication skills, classroom management, handling laboratory activities and reinforcing their learning. Not only did pupils have views on these different dimensions of effectiveness which they were quite happy to articulate through the questionnaire, but these views coincided very closely with those of the teacher mentors and university tutor who were supervising the student teachers. The pupils were quite able to discriminate between most and least effective teachers. I will return to this point later in considering how such information could be used.

Becoming effective: the start

If you are training to become a teacher in England and Wales you will be aware that your PGCE 'year' is largely spent in school. While you are doing this, clearly not all of the time will be spent in the science laboratory. There are many other things that you will experience and many people with whom you will talk, besides your new science colleagues, in order to begin to appreciate the whole sweep of the dimensions of the teacher's role. Nevertheless, your main reason for being there is to learn how to teach. There is a sense in which no one can teach you how to teach, you have to learn it for yourself. An analogy is that learning to teach is a little like learning to ride a bicycle. You do it by attempting it. Of course it is very useful (even vital) to have someone to keep you upright and to prevent you from falling off. And the initial attempt at riding must be a very short one! Your first attempt at teaching does not happen the first time that you visit a school. You will find that your school-based programme is carefully structured so that you make that first attempt when you, and everyone else concerned, are confident that you are prepared for it. The structure of your programme may well look something like this:

Weeks 1 and 2:	Some structured observation of one or two classes, accompanied by opportunities and experiences to allow you to find out about routines and procedures in the science department.
Weeks 3–6:	Working within classrooms, helping with group activities, talking to pupils during practical work etc. Undertaking responsibility for planning and giving part of a lesson e.g. the introduction or the conclusion.
Weeks 6–8:	Extending responsibilities by some team-teaching.
Week?	Being responsible for planning and teaching a whole lesson.

Our students at Hull undertake a series of activities by which they familiarize themselves with the school and the science department in the early stages of their placement. They: read and discuss departmental schemes of work; examine resources such as textbooks, video, reprographic and IT facilities; work alongside a laboratory technician for a day; find out about the roles of colleagues with specific responsibilities such as KS3 coordination, assessment, IT, special educational needs; shadow a pupil for a day, shadow a member of the science department, including a support teacher, etc. They also spend time observing classes. This requires more extensive comment.

Observing classes

The reason for suggesting that you might spend time watching other teachers teach is obvious – you will learn something from it. But what exactly do we mean by that? Certainly you will see what teachers and pupils do in lessons and so you will become familiar with these events. In the same way, one can watch a play or a film and become familiar with the story, the characters, the scenery and so on. By doing so, however, it is most unlikely that the theatre or cinema-goer becomes able to act!

And yet this seems to be the assumption when you are asked to observe teaching – at some time later, you will be able to put into practice what you have seen taking place. So, the argument goes, the more effective is the teaching you observe, the better you will be placed to put that into effect. There are a number of problems here which must be resolved. First, the complexity and subtlety of the process can be bewildering. Good teachers have an almost chameleon-like facility to accommodate to the change in mood, responsiveness and needs of a class of pupils. And what is even more puzzling is that they do this in very different ways. Since how they operate is, essentially, part of their personality, good teaching is difficult to analyse and almost impossible to imitate. There is certainly not just one type of effective teaching. The second difficulty lies in the overall notion of the effective teacher. Perhaps a more appropriate focus is not on the person but on the act, that is what aspects of teaching were performed effectively. This is likely to be a useful idea since the level of effectiveness is bound to vary throughout the lesson. An effective introduction is not always followed by a good summary, or effective organization of materials does not guarantee an effective question-and-answer interaction.

It is essential then to undertake your observation in a way that allows you to analyse what is happening. In order to do this you should concentrate on particular aspects or phases of the lesson. You should also do this for some particular purpose, e.g. writing a short report of your observations, preparing for a piece of teaching that you are to give in the near future or to provide appropriate background information for a task which you might set yourself or which is set for you such as writing a retrospective alternative introduction to the lesson or running a safety check on the materials used. In other words your *own* learning needs to be an active process.

A very wide range of suggestions for observing different aspects of classrooms is available;[5] some examples which are appropriate for the early stages of your classroom experiences are given below.

1. Use a checklist to give you some idea of the rules and routines of the class, e.g. are the following standard practice for pupils:
   (a) stand up at the beginning of a lesson;
   (b) choose where to sit;
   (c) go to allotted spaces;
   (d) work in self-selected groups;
   (e) help each other in their work;
   (f) expect not to consult each other;
   (g) put hand up before speaking to the teacher;
   (h) speak directly to the teacher, butting in at any time;
   (i) be silent when the teacher begins to speak;
   (j) carry on with what they are doing when the teacher speaks;
   (k) move about freely during a lesson;
   (l) compose their own notes;
   (m) copy notes from the board;
   (n) be expected to have with them pencils, rulers, rubbers, etc;
   (o) be allowed to borrow these items;
   (p) be allowed, if they finish early, to get on with homework.

2. How was the lesson introduced? Did the teacher use an event, object or process to gain motivation (or was it 'Can anyone remember what we did last lesson?'). How was pupils' involvement gained at the start? What was the mood and behaviour of the class (excitement, enthusiasm, boredom, resignation)? How long did it take for the class to settle? Which pupils took longest and why? Did the teacher start to speak before the class were all attending?

3. What did pupils actually do in the lesson? How long did they spend on each part of it (e.g. listening, writing, doing practical, clearing away)? How much time did they spend off-task? How did this vary for particular pupils?

4. How were the pupils briefed for the various phases of the lesson (worksheet, blackboard, overhead projector)? Did the briefing involve taking pupils through worksheets, etc? Were pupils briefed for one activity at a time or for a sequence? How did the teacher check that everyone knew what to do? How were transitions between different activities handled?

5. For a practical lesson: Where do pupils put coats, bags, etc.? How were the pupils grouped? How were equipment and materials distributed? What safety reminders were given and how? What did

the teacher *do* during the practical? What provision was made for pupils who worked at different rates? How was clearing up/washing up done? How did pupils conduct their written tasks, e.g. were they briefed about how to record observations and write up their work?

6. How many pupils were addressed by name? Were questions asked of individuals or directed at the whole class? Were more questions addressed to boys or to girls? Were questions directed to all parts of the laboratory? Were only pupils with their hands up asked questions? Were questions open or closed ('What do you notice about . . .?' as opposed to 'What's the width of a red blood cell?')? How many questions probed recall of an earlier fact or event; encouraged observation; or encouraged thought and helped reasoning?
7. How was the involvement of pupils encouraged? Were there instances of verbal encouragement ('Good . . . Well done . . . That's right . . .')? What examples were there of non-verbal encouragement, e.g. a smile or a nod? How were wrong answers received and dealt with?
8. How did the teacher get silence at the start of the lesson? During the practical phase? What control measures were employed? What instances were there of the use of sanctions to deal with minor misdemeanours? Verbal and non-verbal? How was eye-contact used? Tone of voice?
9. How was the lesson concluded? Were key points reinforced? How was continuation for next time indicated? What type of homework (if any) was set? How were pupils dismissed from the lab?

Some suggestions for specific follow-up activities having observed a lesson might include:

1. checking where that lesson fits into the departmental scheme of work;
2. speaking with the teacher about the National Curriculum targets that were involved in the lesson;
3. writing a lesson plan for the lesson which you review with the teacher;
4. finding out how pupils will receive feedback on their written work by helping to mark their books;
5. designing some activities that meet special educational needs in that lesson, e.g. for a very able pupil or a partially sighted one;
6. analysing a tape-recording of the lesson (taken after obtaining permission to do so!) for different aspects of language use, e.g. how technical terms were explained;

7. finding out how the COSHH assessment had been done on the lesson;
8. thinking about potential use of audio-visual or IT resources for the lesson if neither sort has been utilized.

Your observation of teaching will provide you with many insights into classroom life and will give you many ideas that you will be able to try out in your own teaching. Before moving on, however, a word of caution about your observation. On some occasions you will see an experienced teacher perform with seemingly effortless brilliance, without a lesson plan in sight. You will feel that you will never be able to teach like that! At other times you may see the same teacher pacing himself or herself between the hours of 9 and 4, probably so that they can mark books until 12 midnight. They will be employing a range of coping strategies – energy-saving and stress-reducing tactics. It is vitally important that you learn to distinguish between these different facets of teaching. If you confuse the two and model your own teaching on the coping strategies, obviously you will never attain that state of effortless brilliance. Experienced teachers have become so by undergoing a developmental process, each step of which has involved mastering techniques such as how to plan lessons, how to use resources effectively and how to monitor their progress.

First lessons and subsequent progress

Despite the fact that your introductory experiences will have been designed to allow you to settle down and to feel comfortable with pupils, PGCE students and new teachers always comment on the high degree of tension they experience in standing in front of a new class. Typically you will be concerned with what you are going to say and do. Will the pupils like you? Will they do what you want them to do? What will you do if they do not stop talking? That is, all your concerns are teacher-based where the teacher is you! Gradually you will find that your concerns will shift to thinking about the pupils. What do you want them to do? Have they understood what was expected? Do you need to go over anything again? Do you need to give another example, to approach the idea in a slightly different way?

Essential to the process of developing insights into pupils' needs is your ability to evaluate your work. This needs to be done in a structured way at the start. You might try to identify (and discuss with your supervising tutor or another colleague) one or two aspects of a lesson that seemed to go well and one or two that went less well, and the reasons involved. This point is developed in more detail in Chapter 3. As your experience widens such evaluations should develop into a deeper and more extensive process of reflection on your progress. This might take the form of a diary or a journal. It could be structured around key aspects of the teacher's role such as the more searching competences identified by the DfE. When you feel secure in your teaching, the process of reflection can be made very much more meaningful by using the feedback that can be gained from pupils about the lessons or from the presence of another colleague.[6] Current developments involving records of achievement in schools provide an excellent vehicle for your to begin to explore pupils' perceptions of your teaching. Some examples of how more detailed feedback can be obtained from pupils are to be found in the references listed at the end of this chapter.[7]

Effective biology teaching

This book is about the effective teaching of biology. Most of the points that have been raised and discussed earlier might apply to any area of the curriculum. Clearly, teaching biology well rests on sound general principles of effective teaching but, additionally, it carries unique possibilities and demands because of the subject matter with which it is concerned. We could deal at some length with philosophical ideas about the nature of biology, what distinguishes it from other sciences (taken further in Chapter 2) and why we consider it to be a subject in its own right. I do not intend to do so here except to say that what characterizes it is that it involves the study of living organisms. Much of the terrain involved is mapped out in the chapters that follow.

Teaching and learning of biology has been changed in England and Wales with the advent of the National Curriculum in at least two ways:

1. all pupils will learn biology up to the age of 16, either as a separate subject or as part of a programme of balanced science;
2. most primary and secondary science teachers irrespective of their subject disciplines will teach aspects of biology to some pupils.

Indeed one of my main justifications for writing this book is because more pupils will study biology and a wider range of teachers will teach it than ever before.

It is interesting to speculate why we teach biology, or any other subject for that matter, at all. Why should particular subjects occupy the place they do in the curriculum? Of course the status of subjects in the school curriculum varies enormously with place and time as competing pressures of different viewpoints jostle for a share of curriculum time. The question of the inclusion of biology in the curriculum is often raised when interviewing prospective candidates for PGCE courses or for first appointments. Surprisingly, few people seem to have considered it. When pressed, points about the importance of understanding societal issues emerge such as pollution and the environment, organ transplantation, IVF, HIV/AIDS, etc. An interesting piece of research was conducted a few years ago by Arthur Lucas. He was interested in finding out what the general public knew about some straightforward ideas in biology. So he set some test questions to a representative sample of over 1000 members of the British public over the age of 15. The demands of the questions were about those expected of average GCSE candidates. The findings make interesting reading: for example about 50% knew that the sex of a baby is determined by the sperm not the egg, 52% knew that genes were responsible for ensuring that offspring were similar in many ways to their parents and just under 40% thought that freezing meat killed harmful bacteria. Lucas concluded that we should keep such findings in mind in our curriculum discussions:

> There is a danger in so emphasising the modern, the esoteric, and the economically applicable that we forget that there is still a need for the mundane, the basic and perhaps even those topics that have become boring to teach.[8]

Notes and references

1. Department for Education (1992) *Initial Teacher Training (Secondary Phase), Circular 9/92 Annex A.*
2. Office for Standards in Education (1993) *Working papers for the inspection of secondary initial teacher training*, p. 25.
3. Treagust, D. F. (1991) A case study of two exemplary biology teachers. *Journal of Research in Science Teaching*, vol. 28 no. 4, 329–42.
4. Tairab, H. H. and Wilkinson, W. J. (1991) Evaluation of biology trainee teachers' effectiveness using their students' perceptions of teaching behaviour. *Research in Science and Technological Education*, vol. 9, no. 1, 51–62.
5. See for example Perrott, E. (1982) *Effective Teaching: a Practical Guide to Improving your Teaching*. Harlow, Longman.
6. Procedures for appraising teachers are now in place in schools. The aim of such procedures is professional development.
7. See Baird, J. R., Fensham, P. J., Gunstone, R. F. and White, R. T. (1991) The importance of reflection in improving science teaching and learning. *Journal of Research in Science Teaching* vol. 28 no. 2, 163–82. Reprinted in Whitelegg, E., Thomas, J. and Tresman, S. (eds) (1993) *Challenges and Opportunities for Science Education*, pp. 61–78. London: Paul Chapman and Open University Press.
8. Lucas, A. (1987) Public knowledge of biology. *Journal of Biological Education*, vol. 2, no. 1, 45.

CHAPTER 2

Biology in the science curriculum

There has been no better statement of the purpose and conduct of teaching science in school than that provided by the Secondary Science Curriculum Review.[1] Among sixteen recommendations for achieving 'science for all' (an aim in itself) we find:

> The science studies which are undertaken should be broad, in the sense that they should include elements of astronomy, biology, chemistry, computer science, earth science, electronics, physics and technology. The approach to science studies should display structural coherence, continuity and balance.[2]

Biology thus makes an important contribution to the provision of breadth, balance and coherence as a component of a science curriculum. That contribution is possible because of the differences between biology and the other sciences. Writing in a different context, Shayer and Adey[3] have pointed out that because the sciences are different, their teaching demands different approaches. They argue that while the methods of experimental design in biology are sophisticated in their demands, the concepts that are involved are capable of being understood at a number of different levels. Because of this they claim that:

> . . . there is a complex web of inter-relationships between different topics at low, intermediate and higher levels, and therefore many different starting points and sequences which are possible, and many different ways of dividing the subject into areas of study.

However, when you take up your first teaching post you will not find yourself in a situation where you are free to teach whatever aspects of biology you wish in whatever order or way you would like. You will be guided by the requirements of the National Curriculum (if you work in England or Wales), the general curriculum aims of your school, the examination syllabuses followed by your science department and the schemes of work developed by your colleagues in school. That is not to say that the new teacher in school is expected to behave like a puppet; plenty of opportunities will exist for you to show personal initiative. One of the characteristics of the job of teaching has always been the freedom with which ideas can be advanced and shared with colleagues with whom you work on an equal footing.

Despite what was suggested earlier about how biology is different from other sciences, teaching a biology topic involves considering the same fundamental issues as arise in teaching topics in chemistry, physics or whatever. Two central questions are: *what* are you going to teach (what will be the outcomes of the lesson, what will the pupils gain as a result of it?) and *how* are you going to do it (what teaching methods will you employ?)? Not only are these questions important individually but also the balance and interaction between them are crucial. And this takes us to a brief initial consideration of the National Curriculum in science.

Science in the National Curriculum

When you examine your new teaching timetable for the first time you may well find that the subject biology is not on it. This should not surprise you; you will have been prepared for this during your training course. Biology is there of course although it may be subsumed under the title 'science' even up to and including the GCSE classes. At the time of writing, schools in England and Wales are in the process of deciding how they will organize the science curriculum at key stage 4. In September 1992 children in most schools in England and Wales embarked upon new GCSE syllabuses in science teaching either to the Science: Single Award or to Science: Double Award. As a new teacher in school you will join a department that will have made the decision about which of the available GCSE syllabuses is to be followed. Indeed, you may find that your department has decided to follow the GCSE courses in the separate sciences which will be examined for the first time in 1995.[4] All GCSE syllabuses conform with the National Curriculum Order for Science: Double Award schemes, the biology coverage is specified by Attainment Target 2: Life and living processes and the corresponding sections of the programme of study. GCSE biology syllabuses are augmented by a 25% subject extension, determined by each GCSE group. In this way, the outcomes of the work in terms of the *knowledge* and *understanding* which pupils are expected to gain from it are specified. Hence at one level, albeit a superficial one, the answer to the question stated earlier, what are you going to teach, is what it says in the relevant sections of the National Curriculum. And the same answer, but relating to Attainment Targets 3 and 4, could be given to you for topics in physics or chemistry. That answer is unsatisfactory as far as detailed lesson planning is concerned and it is one of the functions of a departmental scheme of work to indicate how the topics in the programmes of study should be approached to achieve the outcomes detailed in the statements of attainment.

Attainment Target 1: Scientific investigation is different:

> Pupils should develop the intellectual and practical skills which will allow them to explore and investigate the world of science and develop a fuller understanding of scientific phenomena, the nature of the theories explaining these, and the procedures of scientific investigation. This should take place through activities that require a progressively more systematic and quantified approach which develops and draws upon an increasing knowledge and understanding of science. The activities should encourage the ability to plan and carry out investigations in which pupils:
> i) ask questions, predict and hypothesise;
> ii) observe, measure and manipulate variables;
> iii) interpret their results and evaluate scientific evidence.[5]

This identifies the need for pupils to carry out activities in which they will develop investigative skills and an understanding of the nature of science. The other question posed earlier, about the teaching methods to be employed, is partly answered by the activity-orientated, investigative nature of Attainment Target 1. The relationship between Sc 1 and the other attainment targets (ATs) is important:

> Through investigative activities in science pupils are able to develop their ways of thinking about science and their understanding of the concepts and knowledge in attainment targets 2, 3 and 4.

and

> Work related to Sc1 cannot be tackled in isolation from the rest of the programmes of study. All investigations which allow pupils to show their achievements in Sc1 should be based in a context related to one or more of the other attainment targets.[6]

Hence Sc1 forms the context and approach (the 'methods') of achieving *some* of the knowledge and understanding outcomes of the other ATs. Put another way, Sc1 does not exist on its own independently of the other ATs, it comes into being within the work designed to address them. Clearly this is an important and complex area and I will return to it later, but at this stage it is important to be clear that it is *not* true to equate Sc1 with practical work and the other ATs with 'theory'. As we will see later that distinction is not particularly useful. Sc1 is about doing investigations which, with other sorts of 'practical' work, are involved in teaching Sc2, 3 and 4. As a science teacher (at whatever key stage) you will be trying to have your pupils attain the objectives specified as statements of attainment in Sc2, 3 and 4 and you will be doing so through the approach of Sc1 (and hence attaining its specified objectives). Thus it is the *commonality* of the approach that is important, one which stresses the essentially similar nature of 'doing science' whether it is biology, chemistry or physics. It is in that way that the pupils will become familiar with conducting investigations and become progressively more adept at doing so.

At key stages 3 and 4, Sc1 has a weighting of 25% as do the other ATs.[7] Weighting in this sense refers to the proportion of teaching time that should be devoted to the attainment target. It is also the amount of emphasis to be given to it in the associated assessment arrangements. Clearly to be fair to your pupils your teaching must reflect the weightings identified by the National Curriculum. Making decisions about issues of this sort is not easy. Although the National Curriculum provides examples of work that addresses statements of attainment at the various levels in each attainment target, such examples are limited to one or two aspects of the statement and they are not substitutes for it.[8] For example, for Sc1 level 3a 'suggest questions, ideas and predictions, based on everyday experience, which can be tested', the example quoted is 'suggest what is needed to help cress to grow, how these ideas might be tested and what might happen'. This example is drawn from Sc2 and clearly connects with Sc2, 3c 'know that green plants need light to stay alive and healthy'. But, the SoA from Sc1 quoted above could be tackled in the context of, say, Sc3 3c: 'understand some of the effects of weathering on buildings and rocks'. The latter SoA does not *have* to be addressed by an investigation, but it could be if you wished. The methods of conducting investigations using cress seeds or on weathering of buildings and rocks will be similar in that a question must be posed in each case, observations and measurements might be made, conclusions would be drawn and explanations offered. In other words, the investigative skills or processes are common to both situations even though dark boxes, different thicknesses of tissue paper etc. are used in one lesson and test-tubes, balances and, perhaps, dilute acids are used in the other. You might decide that it is appropriate to tackle both areas of knowledge and understanding by pupil investigations, that decision is yours (or your departments). It is clearly related to the idea of balance in your teaching: every area of work cannot be used for investigations of the sort demanded by Sc1 but enough work must be treated in this way to meet the weighting requirement mentioned earlier.

'Processes' in science teaching

It is necessary to decide on an appropriate balance to aim for in your teaching between the emphasis you give to learning facts and concepts on the one hand, and to gaining expertise in ways of doing science, the so-called 'processes' of science on the other. The process–content debate has been an issue in science education for some time.[9] Over the last twenty years or so an increased emphasis has been given to teaching processes and skills, perhaps with biology teachers fighting something of a rearguard action.[10] Entire curriculum projects have been designed round lists of processes. *Warwick Process Science*[11] identifies the following: observing, inferring, classifying, predicting, controlling variables and hypothesizing. *Science in Process*, the very popular ILEA package,[12] is based on: applying, interpreting, classifying, investigating, evaluating, observing, experimenting, predicting, hypothesizing, raising questions and inferring. The key point about these, and similar, curriculum development packages (which predate the National Curriculum[13]) is the intention that these processes are developed through the teaching by making them central to each classroom activity. They are also made explicit to the pupils by encouraging pupils to focus on them as outcomes.

Aspects of this approach have resulted in extensive criticism, most particularly that:

1. it is an inaccurate representation of the methods of science;
2. the processes are not confined to science, but they are general cognitive skills that are routinely employed throughout life without the need for formal instruction in them;
3. there is no evidence that performance at them is improved by repetition.[14]

Many teachers also report problems in handling the language of processes with eleven and twelve-year-old pupils and hence in communicating to them what the aims are for a piece of work. Despite these criticisms these curriculum materials are worth examining when you are trying to identify areas for inclusion in your lessons. They illustrate very well that an activity can be structured in different ways to achieve different outcomes. And many of the outcomes listed in the National Curriculum in Sc1 are process-based ones.

Further principles for selecting teaching material

The question of content is a minefield. One person's clutter is another person's gem. Perhaps it is more important that as teachers we are satisfied with our reasons for choosing items of content and also we can then justify our choices to others, e.g. the children.[15]

Different people will have different reasons for wishing to include some things and to exclude others from their schemes of work. We all feel safe with what is familiar to us. Whether this should form the basis for selecting what to teach our pupils is another matter. It happens! When option courses are selected in modular science programmes at GCSE or at A level, it is not unknown for the luxury of choice to be exercised by the teacher rather than the taught! Hence at A level microbiologists choose the microbes option, geneticists the applied genetics and all for the best of all motives: being able to teach it more effectively.

The Secondary Science Curriculum Review proposed the following criteria for selecting content in school science:

1. The content included should maximize the possibility of teaching the processes, skills and attitudes important to science, and increase the competence and capability of pupils.
2. It must be able to be taught to, and owned by, all pupils.
3. The content should have some relevance and applicability. It must need to be known and be transferable to a range of contexts during and after school.
4. The content must appeal to both male and female students, and to students of differing cultures.
5. Science content should be consistent with a broad and balanced science curriculum.
6. The necessary teaching strategies and resources must exist to teach it.
7. It must provide a basis for further study.
8. It must take into account local circumstances.[16]

The general aim of the Review was to make school science more relevant and applicable to pupils than it was at that time. These are useful ideas to have in mind whether it is a curriculum being planned, a scheme of work for a year group or even an individual lesson. Of course 'relevance' has a number of different dimensions. Social relevance involves notions of preparing pupils for their future roles in a democratic society. Industrial relevance might involve providing information about industry and how it makes use of scientific ideas as well as in developing the skills needed for future employment.[17] Much more important, however, is that the curriculum should be *personally* relevant to pupils:

> Perceiving personal relevance in a science topic is likely to engender in pupils curiosity, interest, enjoyment, enthusiasm and a sense of satisfaction.

and

> Students are more likely to learn some new ideas and skills if they see some point and personal significance in doing so.[18]

It is in this area of personal relevance that biology as a science enjoys enormous potential advantages over other areas of the curriculum. Children are drawn to living things from a very early age, they are naturally curious about them, particularly animals (plants less so, a point taken up in Chapter 5), and they want to find out more about them. As they get older, pupils wish to find out more about themselves, a natural curiosity that can be exploited. Nevertheless teaching some areas of human biology to adolescents has some pretty obvious pitfalls and poses a challenge for the teacher as great as one might encounter anywhere in the curriculum.

Realizing the vital importance of building on the principle of personal relevance has come as a result of our awareness that there has been a decline in the popularity of science over the last thirty years. We also know that interest in science deteriorates in individual pupils as part of a general process of declining motivation towards school and teachers as pupils get older.[19] The unpopularity of the sciences, particularly physics, is reflected in declining take-up of the subjects at A level and therefore in degree courses. Building a national curriculum in science that is based on the principles of balance and breadth across the three main sciences is a bold step! It remains to be seen how this will affect pupils' attitudes to science and their desire to study it further.

A question being asked currently is: Will pupils who have studied balanced science (in 20 per cent of the weekly curriculum) 'know' enough to follow A-level courses in the separate sciences? However, for several reasons, this is the wrong question to ask. Advanced level courses are not immutable, work is in progress to bring them into a position in which they build on the National Curriculum. More importantly, will pupils who have studied science up to sixteen *want* to go on to further studies in it? By imposing elements of all three main sciences on pupils, it is possible that they will not only reject the physical sciences as they have done in the past (particularly girls), but also biology which up to now has retained the interest of adolescents more effectively. Reviewing the requirements for a new curriculum that might offer greater appeal to adolescents, John Head comments:

> Not only can a more central role for biology be envisaged in a new curriculum but some introduction to psychology and anthropology might be made. Possibly the balance should be that physical sciences take no more than half the time in the schools science curriculum with the other half devoted to the biological and humanistic aspects.[20]

He goes on to say that science is made meaningful for pupils (i.e. relevant) when teachers employ the following tactics:

1. Give pupils opportunities to reformulate concepts in their own words whether it is during written work or in discussion in small groups as in practical work.
2. Have pupils use newly learnt concepts in novel situations. This idea finds expression particularly, but not exclusively, in investigative work.

3. Allow the learner to make use of the prior knowledge that he or she brings to the situation.
4. Involve the learner in the teaching strategy by sharing with the teacher a knowledge of the purpose of a particular topic or activity.

How these tactics can be employed in teaching biology topics is taken up in subsequent chapters.

Notes and references

1. The materials published by the Secondary Science Curriculum Review are essential reading for all new science teachers. The recommendations referred to here are in *Better Science: Key proposals. Curriculum Guide 1.* London: Heinemann Educational Books, 1987.
2. SSCR op. cit. p.4.
3. Shayer, M. and Adey, P. (1981) *Towards a Science of Science Teaching.* London: Heinemann Educational Books.
4. The philosophical and historical twists and turns by which the current situation has arisen are outside the scope of this chapter. If you are interested in this, talk with your curriculum-scarred, experienced colleagues.
5. DES/WO (1991) *Science in the National Curriculum*, p. 14. London: HMSO.
6. SEAC (1993) Assessing Sc1 in *School Assessment Folder.*
7. DES (1991) *National Curriculum: Mathematics and Science Orders under section 4, Circular No 17/91*, p. 5.
8. NCC (1991) *Science Non-Statutory Guidance.*
9. For an account and analysis see Wellington, J. (1989) (ed.) *Skills and Processes in Science Education, a Critical Analysis.* London: Routledge.
10. An analysis of the teaching styles of science teachers showed that biology teachers were more likely to adopt a fact-orientated non-practical approach than their physics and chemistry colleagues. See Eggleston, J. F., Galton, M. and Jones, M. E. (1976) *Processes and Products of Science Teaching.* London: Macmillan Education.
11. Screen, P. (1986) *Warwick Process Science.* Southampton: Ashford Press.
12. ILEA (1987) *Science in Process*, Ten Units and Teachers' Guide. London: Heinemann Educational Books.
13. Many of these materials were published in response to DES (1985) *Science 5–16: A Statement of Policy*, London: HMSO. Some are currently under revision to meet the requirements of the National Curriculum.
14. Millar, R. (1989) 'What is scientific method and can it be taught? in Wellington, J. (ed.) op. cit., pp. 47–62.
15. SSCR (1987) *Better Science: Choosing Content. Curriculum Guide 2*, p. 9. London: Heinemann Educational Books.

16. SSCR (1987) ibid. p. 10.
17. The last aspect is much less important because most of the skills required will be specific rather than generic and hence taught by industry itself.
18. SSCR (1987) *Better Science: Making it Relevant to Young People. Curriculum Guide 3*, p.4. London: Heinemann Educational Books.
19. Head, J. (1985) *The Personal Response to Science*. Cambridge: Cambridge University Press.
20. Ibid. p. 85.

CHAPTER 3

Teaching with a practical emphasis

Practical work – some clarification

The nature, purpose, use and abuse of practical work in the school science curriculum have received constant comment and debate for at least the last thirty years! Research has identified the following five major sets of reasons offered in support of laboratory-based work:

1. Science involves highly complex and abstract subject matter. Many students fail to comprehend such concepts without the concrete props and opportunities for manipulation afforded in the laboratory.
2. Students' participation in actual investigations developing their enquiry and intellectual skills is an important component of learning science as enquiry.
3. Practical experiences, whether manipulative or intellectual, are qualitatively different from other experiences and are essential for the development of skills and strategies with a wide range of generalizable effects.
4. The laboratory has been found to offer unique contexts for the identification, diagnosis and remediation of students' misconceptions.
5. Students usually enjoy activities and practical work and when offered a chance to experience meaningful and non-trivial experiences they become motivated and interested in science.[1]

However, fairly trenchant criticism of the nature of the practical work done in schools has also been offered:

> Much that we provide is ill-conceived, muddled and lacking in real educational value.[2]

Rosalind Driver maintained that:

> The slogan 'I do and I understand' commonly used in support of practical work in science teaching might better be replaced by 'I do and I am even more confused'[3]

Dorothy Dallas advises that

> teachers need to look critically at proposed practical work – is it merely recipe-following, without comprehension, the . . . equivalent of sand and water play in the nursery school? Or are the class really learning from it?[4]

Even defining practical work is not easy. After re-affirming recently the central place of practical work in school science SEAC categorize practical activities into four broad types: basic skills exercises, observation exercises, illustrative work and investigations.[5] Apart from investigations, the other types of practical work are not defined, explained or exemplified. Rather more helpfully, Woolnough and Allsop suggest that practical work should involve *exercises*, by which practical skills and techniques are developed, *investigations*, based on problem-solving activities and *experiences*, by which pupils 'get a feel for phenomena'.[6]

Putting this in the context of biology, the purposes of practical work (when resources are used in activities performed by small groups of pupils) are as follows:

1. Developing an understanding of:
   (a) structures, e.g. lungs, heart, leaves;
   (b) processes *in* living organisms, e.g. digestion, respiration.
2. Developing skills in:
   (a) manipulation, e.g. measuring out small volumes, making temporary slides;
   (b) the processes *of* science, e.g. observing, recording, interpreting, hypothesizing, etc.;
   (c) conducting whole scientific investigations, e.g. what affects the growth of duckweed?

It is well to remember that we have a wide range of teaching methods and resources at our disposal, of which small-group practical work is one option. Much more important is the question of whether the pupils are actively engaged in their learning. On some occasions alternatives to practical work may be more effective. Hence the decision to undertake a piece of practical work should be made in the light of knowing the full range of resources available such as video, games, simulations, role play, CAL, written material, discussion, debate or even demonstration by the teacher. So, if different types of practical work and other learning experiences can be used to fulfil rather different purposes, it is essential that you think clearly about what you wish pupils to gain from your lessons and then select the most appropriate means of achieving that. And that means careful planning and effective management, not only of pupils, time, books (and other written resources) but also of apparatus, chemicals and the biological materials involved.

Planning practical work

A scheme of work is a device by which a curriculum plan, a course of study or an examination syllabus is converted into the operational lesson-by-lesson intentions by which teachers will bring about pupils' learning.

Such a scheme of work is a departmental document, it will have been produced by members of the department (usually having been heavily influenced by the Head of Department) and is intended for use within that department.

A scheme of work should contain the following:

1. An indication of how the work addresses the general aims of the school and conforms with or exemplifies school policies, e.g. on assessment, recording and reporting.
2. A list of the topics to be covered, the order in which they are to be covered and their respective time allocations.
3. For each topic, ideas for a sequence of lessons, the pupil outcomes expected, the teaching approaches recommended and resources available, e.g. video, IT, practical materials, etc.
4. Guidelines on assessment procedures: When? How? What? What use is to be made of the information?

Such a document will be upgraded periodically in the light of on-going evaluation of areas 2, 3 and 4 by the department. If you work in a department with a scheme of work like this, congratulations! That is the ideal situation: in practice you are likely to find your scheme of work involves a few cursory headings scribbled down from the programmes of study from the National Curriculum, e.g. Year 9: metals, air, sound, cells, reproduction, etc. Alternatively your new science department may be using one of the published schemes (which may even predate the revised National Curriculum order). It is easy to allow yourself to be lulled into thinking that your planning has been done for you. It hasn't! Such a thought can be positively dangerous. What you will have before you is someone else's ideas. These will not necessarily be appropriate for you and your classes in your particular situation. On the other hand, these materials will usually have been written by experienced teachers and can form a valuable starting point for you. Your greatest concern at this stage is likely to be 'what shall I do with the pupils?', i.e. how are you going to keep them gainfully occupied for sixty to seventy minute lessons? Hence the content of such published schemes, along with other sources of ideas to be found on the departmental bookshelf, will be useful.

Once an overall sequence of lessons has been decided you will turn to the detailed planning of each lesson. Before doing so the sequence should be rechecked: for its appropriateness in meeting the content and time demands of the topic, to ensure that there is a genuine developmental sequence in the topic in that later ideas build on earlier ones; so that a variety of activities are experienced by the pupils and that the resource demands can be met. Your department will have its own procedures for organizing resources needed for individual lessons. Most departments operate a system by which laboratory technicians or assistants are briefed about requirements up to a week in advance of the lessons concerned. How this system operates will be one of the first things that you will need to find out about on joining the department. You will in all probability be briefed about this and procedures for using video, IT, etc.

However, the resources you will need for your lessons are your own responsibility. Obviously then as part of your planning for the topic, you will have decided what resources you want. In the case of lab-packs or chemicals this is fairly straightforward. For biological material it may be far less so. The sort of resources you will need are things like pond water, *Daphnia* (water fleas) privet leaves, *Elodea* (Canadian pond weed), duckweed, snails, woodlice, dandelion fruits ('parachutes'), etc. Knowing what to use for your lessons and how to get this sort of material will become part of your resource bank of ideas, which will develop throughout the whole of your teaching career. Such ideas should be collected in a file or on disk from the moment you decide to become a teacher. They are gleaned from a wide variety of sources and are essential to you in your planning and teaching. Becoming a good teacher of biology depends to a considerable extent on being able to call upon such a bank of ideas for lesson content.

There is a difference between knowing what you want for your lessons and knowing how to get the resources. Many such resources are to be found in the schools' surroundings, your garden (help, I live in a third-storey flat without even a window-box!), the local park, pond or wood. Who collects these for you? You may have highly experienced, professional laboratory technicians who are not so run off their feet in school that they have time to do the collecting for you! Or you could be lucky in that your pupils will remember that you told each of them to bring in ten woodlice for tomorrow's lesson. But don't bank on it! That leaves you, but don't worry, the time which is spent on collecting such living material is more than satisfactorily repaid in terms of the pupils' responses to it in your lessons. What if you don't know where to find these resources? Clearly that is what colleagues are for but remember that most experienced teachers are happy to help those they see being prepared to help themselves. The other golden rule about collecting material from the locality is do not leave it until the last minute – it is surprising how elusive woodlice become if you need to collect 200 of them ten minutes before a lesson! Invertebrates of this sort can usually be collected days or even weeks ahead and maintained satisfactorily in the preparation room until required. Other sorts of biological resources will usually have to be ordered specially from the various supply agencies – procedures with which you will become familiar as a member of the science department.

Lesson Planning

In another volume of this series Elizabeth Perrott[7] has given very sound general advice about how to plan lessons. Our PGCE students in Hull are advised that they need to produce detailed lesson plans for their own use through the PGCE year, and throughout the first year of teaching. Your own ways of planning will change as you become more experienced; however the following checklist may be useful at the start of your teaching.

What does a lesson plan need to contain?

1. Clear statements or objectives for the lesson, having decided the age and general level of ability of the group.
2. A method for knowing that the objectives were in fact achieved (to be assessed in the lesson if possible).
3. The sequence of activities, timed as accurately as possible.
4. The relevant content (knowledge) assumed.
5. The content (knowledge) to be covered during the lesson.
6. The skills required by students during the lesson.
7. The skills to be acquired by students during the lesson.
8. The significant processes involved.
9. The significant student interactions you wish to occur.
10. The kind of report (oral or in writing) the students are expected to produce.
11. The equipment needed.
12. A brief synopsis of what you would say to start the lesson.
13. A brief synopsis of how you would expect to summarize the lesson for the students at the end.
14. Suitable follow-up work – homework perhaps.

A number of aspects of this general strategy are of particular importance.

1. Identifying objectives

Your objectives for the lesson should be framed in terms of specific observable pupil outcomes that you could assess if you wished. They should describe what the pupil will be able to do as a result of the lesson and not what the teacher intends to do.

For example, 'To teach the characteristics of mammals' is not an appropriate objective; 'Pupils will list the following characteristics of mammals: suckle young, have fur, young are born alive' is an appropriate objective. The difference between these is not one of pedantry. The second example provides a clear purpose for the lesson, so that deciding what content and resources to employ follow naturally.

2. Trying it out in advance

This may seem so obvious a point as to insult the reader! When living material is involved only the foolhardy ignore this advice. If you do not try it out first and 'it doesn't work' on the day, all you will be left with is the 'this is what was supposed to happen' excuse. If you have tried it out beforehand and it doesn't work at least you will be able to claim that this demonstrates the variability of living material; give them the results of your previous trial; and escape with some of your credibility intact! Some of the the most straightforward (apparently) and extensively used procedures involving living material contain major pitfalls for the unwary. Here are a few examples.

> Canadian pondweed will not bubble when illuminated unless it is pretreated appropriately: a pinch of sodium hydrogen carbonate added to the water, oxygen bubbled through it and illuminated with a bright lamp (preferably mercury vapour) for a couple of hours beforehand.
>
> Woodlice will not 'behave' properly in choice chambers unless they have been maintained beforehand in a damp environment.
>
> *Pelargonium* leaves retain their starch for at least three days in a dark cupboard – overnight 'destarching' will not do.
>
> It takes ages for pupils to remove chlorophyll from such leaves (a classic example of a potential dead space in the lesson) – use discs punched out with cork-borers instead.
>
> The shelf-life of enzyme preparations varies enormously (even when they are refrigerated), hence your '0.5% amylase solution' cannot be relied upon to produce similar results year after year.

And so on!

3. Introducing it appropriately

How you introduce the lesson is crucial, although the methods by which this occurs will be different for different sorts of lessons. An investigation of the effect of temperature on yeast respiration will be introduced differently from a lesson on the structure of cells.

The introduction is the focusing stage of the lesson during which pupils' attention is directed towards the substance of it and motivation is gained or lost. There is a clear difference between an introduction to the lesson and a briefing for the activity which will follow. The introduction sets the scene. When thinking about introductions for your lessons, try imposing a ban on: 'Who can tell me what we did last lesson?' Instead use some stimulus material such as an object, an event (actual, film, video, audio), a demonstration by the teacher or by making use of a pupil to carry out an activity.

4. Managing the transitions

A lesson sixty or seventy minutes long will contain a variety of tasks and activities for the pupils: listening, small-group discussion, group work, individual writing, etc., perhaps five or six different kinds of activities. Transition from one to another will almost always involve some pupil movement, such as clearing apparatus away, getting out books etc. A great deal of time can be wasted by pupils during these changes. An important principle here is that pupils should always know what they are expected to do next. When carrying out a practical activity they should be briefed about their next task, writing a report, clearing away, etc., before they get started on the activity.

5. Evaluating the lesson

After you have given the lesson, as a matter of course you should jot down your reflections about the success or otherwise of your efforts. Thoughts recorded immediately after completing the lesson will be valuable when you teach the same topic next time.

Try to think about specific events which went smoothly or less so. Comments such as 'the lesson went quite well but the pupils were very noisy' are not useful because they do not focus on the reasons for these outcomes. Evaluations should provide you with a firm idea of what to do next time; of what *not* to do next time or some tactic that might be worth trying. Additionally, these reflections have an important effect in helping you to come to terms with the job. Here is a PGCE student's reflections on one lesson.

Period 2 Integrated Science Year 7

Lesson evaluation: Resusci-Annie doll/First Aid

This lesson was preceded by year 8 filtering experiment and I remember the sink (the only one in this room) is blocked and is being attended to by caretaker and teacher.

Enter (late) pupils, teacher and dummy (the doll!). Coats off, settle down routine, date on board, books, paper, pens, etc. Begin discussion 'What is first aid?' One boy begins definition 'To sustain life, (something else) and promote recovery'. I can't remember the middle bit but its as close to a textbook definition as you can get, what a marvellous start thinks me! Enter another caretaker, followed by headmistress (he pushes me aside, gets a telling off from her later). I carry on, teacher exits.

We discuss various accidents – pupils interested and willing – all. Display Annie, (headmistress and entourage exit). Gather pupils round, dummy does not want to live, refuses to expand chestwise. Nose very stiff, mouth likewise (her, not me). Before blushing occurs, admit failure, allow pupil to try ('I'm not kissing her Miss, she's got lipstick on', oh s--t, I forgot to wipe mine off first). Don't panic, I'm going to clean her first. Of course, according to Murphy's law, she expands (great) other pupils try. At this point, one boy asks another to watch her chest and he replies he would much sooner hold it (in a sexual way). Brief telling-off followed by an anonymous boy's declaration 'Well she's not very big anyway is she?' Explain calmly that this is not acceptable and that if we (the females) refrain from making comments about the inadequacy of their bodily parts we expect them to do the same (and thinks 'besides how the hell would he know about breast size anyway?') This nipped him and the other boys in the bud, so to speak. The boys chorus 'Why doesn't one of the girls try'. Yes, why not indeed, girls feigning shyness (pathetic). I threaten and carry out the choosing of a 'volunteer'. She dutifully obliges and to our amazement not only gets Annie's chest to expand vigorously but also enables Annie's 'wind problem' to alleviate itself. A loud rasping noise emits from said Annie – result we had to hold our sides for 5 full minutes to stop them from bursting. (I shall treasure this cheering-up moment for always.) Even writing this account a few hours later I am still laughing. I then said 'This woman is obviously alive'.

Obtain tissues for those of us who are crying with laughter (I am one of many) and carry on. The verve and vigour of the pupils' willingness to resuscitate Annie after this could be explained in two ways:

1. The camaraderie, ice breaking, enabled them to realize how important the task was.
2. Much more likely, they *all* wanted to make Annie fart, which of course she didn't (these moments are so precious)!

Questions to be asked in setting up practical work

Finally the checklist below has been found to be useful by our students. There are not necessarily any right or wrong answers to these questions but they should be considered and courses of action decided upon as part of the planning process, not during the lesson.

1. What safety procedures/precautions are essential?
   Have you undertaken your COSHH assessment?

2. Demonstration or class practical?
3. How are instructions about procedures to be given?
4. How are pupils to be grouped?
5. How are equipment and materials to be distributed?
6. What are *you* going to do as the practical proceeds?
7. What guidance will have to be given as the practical proceeds?
8. How will pupils be guided to make the observations and measurements to reach the required conclusions?
9. How will observations be recorded?
10. Will there be any dead time?
11. What provision will you make for pupils who are going to work at different rates?
    Will it always be necessary for the slower pupils to finish their work?
12. How is the clearing up/washing up to be done?

Practical activities for different purposes: some examples

Developing understanding of structures

A class has been working on the topic of plant nutrition, the final stage of which involves considering how the leaf is adapted to carry out this process. The aims for this piece of work (not in the form of lesson objectives) are as follows:

1. understanding that gaseous exchange takes place through the leaf surface;
2. appreciating that a leaf is made up of four layers of cells with chloroplasts occurring in the inner two, the outer two being largely transparent;
3. perceiving the presence and significance of a system of air spaces, through which gases diffuse;
4. relating gaseous exchange and light capture to surface area : volume ratios of leaves.

You may find it a useful exercise to try to put these aims into the form of lesson objectives as described earlier.

The teacher decided to use the following activities during the lesson:

1. Pupils submerge a freshly picked leaf (e.g. privet) under hot water (not boiling) in order to observe the appearance of bubbles on the surface of the leaf.
2. Pupils prepare, and examine under the microscope, their own preparations of leaf impressions made in acetate sheet.[8]
3. Pupils mount and examine hand-cut sections of freshly picked leaves prepared in advance by the teacher.

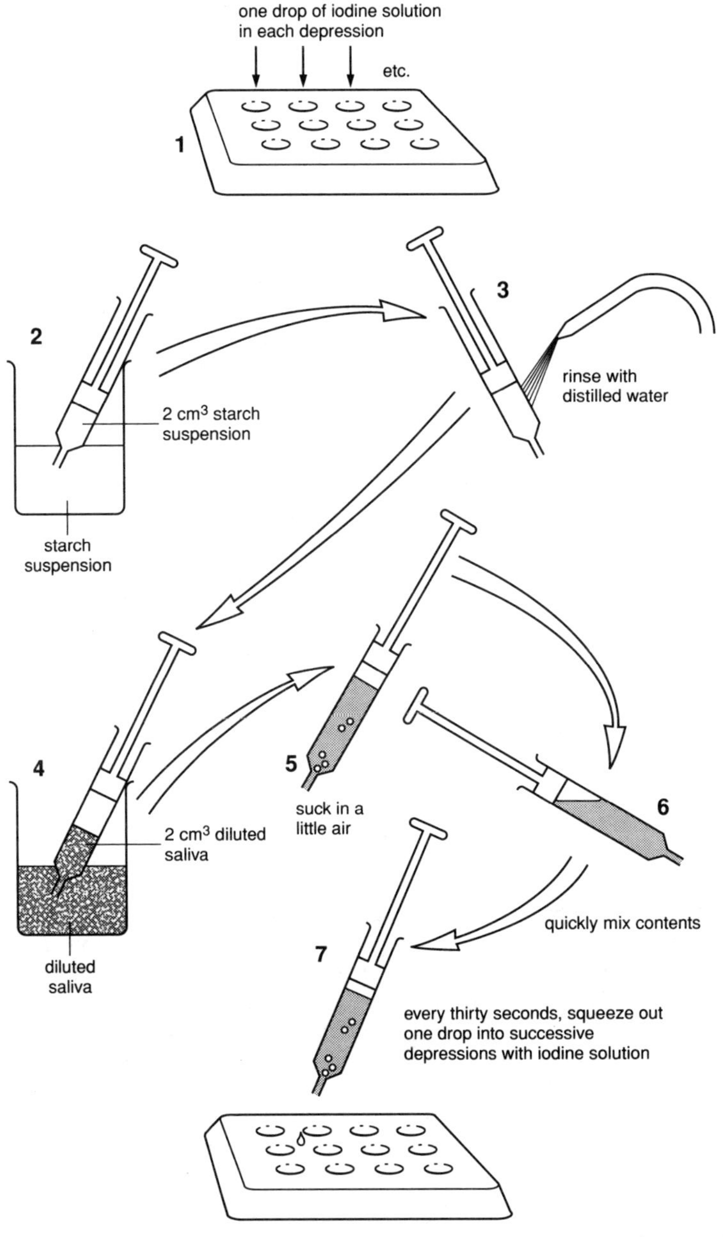

Figure 3.1 An experiment to investigate the action of saliva on starch

4. They examine prepared stained sections of leaves and photographs of leaf surfaces and sections taken under the scanning electron microscope.[9]

Together these activities potentially represent an interesting and effective strategy for achieving the aims stated earlier. However, to be successful the class will need to be briefed thoroughly about each of these procedures (this is not a situation for problem-solving) and the observations made by the pupils will have to be discussed and clarified by the teacher. Further discussion about helping pupils to clarify their observations will be given in Chapter 4, but here is a good example of the role of the teacher in helping pupils to interpret the surface views and sectional views of three-dimensional structures, in this case of cells and stomata.

The teaching sequence, whilst mainly concerned with developing an understanding of structure ('illustrative' and 'observation' exercises in National Curriculum terms), also offers opportunities for pupils to practise manipulative skills (mounting sections on slides, using the microscope, making acetate impressions) as well as process skills of following instructions, observing, recording and interpreting.

Developing understanding of processes in living organisms

The worksheet shown in Fig. 3.1 was followed by pupils working on the topic of digestion. Clearly the main point of this procedure is illustrative, developing understanding of the process of digestion. If the procedure is presented as indicated (i.e. the briefing, how would you *introduce* this activity?) as in the last example, important manipulative skills are also practised, in this case handling apparatus, measuring out small volumes, transferring chemicals safely, etc.

At the end of the procedure, the teacher could ask the class to think about possible reasons for the disappearance of the starch, and so move on to the need to run an appropriate control procedure, to identify further tests to be done on the starch-free product, and so on. In this example, the teacher will have also involved the pupils in process skills, but different ones from those listed above.

So in this way, we see that practical activities are multi-faceted in that they can (and should) address different purposes for doing the activity and justifying the time spent in doing so. Deciding if any practical activity is worth doing thus involves being able to extract maximum mileage out of it by recognizing what it has to offer.

Investigations

In Chapter 2 you saw how the design of the National Curriculum in

Science, with the weighting given to Sc1, imparts an investigative emphasis to the teaching of science. The development of more open-ended work of this sort has been occurring for some time in science courses. Different meanings can be attached to the word 'investigation'. Clearly not all investigations have to be experimental or practical; they might involve a research exercise in a library. However publication of a series of Government documents has formalized what meaning should be given to investigations, at least in the context of the National Curriculum and its assessment. Issues of assessment including those of assessing Sc1 will be taken up in Chapter 8; here we will look at the teaching implications of conducting investigations in the eleven to sixteen age range.

In National Curriculum terms an investigation is an activity in which pupils are encouraged to ask questions, test ideas, predictions or hypotheses and draw conclusions.

Pupils are required to:

1. formulate questions or predictions, using knowledge and understanding at an appropriate level, which they can test;
2. put together a sequence of investigative skills, such as the identification of key variables, observing and measuring into an overall method to answer a question, test an idea or seek an explanation;
3. evaluate their findings in the light of the idea, question, prediction or hypothesis which prompted the investigation;
4. refine and develop the way they carry out investigations.

This involves the process skills of: planning, predicting, hypothesizing, designing, carrying out, interpreting results, evaluating evidence, deducing inferences and communicating plans, tasks, procedures and results.

Put more simply, this involves a three-stage process, corresponding roughly with the three strands of Sc1: *planning*, i.e. deciding what problem to work on; *performing*, i.e. deciding how to tackle the problem and then doing it; *concluding* i.e. deciding what has been learned by carrying out the investigation. Clearly, an essential question to consider is who makes the decisions about each of these stages: pupils, teacher, or pupils and teacher together? The more decisions about each aspect that the pupils make, the more open-ended the activity is for them. In practice each of these aspects can be seen as a continuum, with decisions being made by the teacher, rather than the pupils, moving the activity towards the closed end of the continuum. A further factor that contributes to the openness of the activity is the nature of the problem being investigated: does it have one correct answer or are there a number of potential solutions, each equally acceptable?

The OPENS project[10] gives the following example to illustrate these points. An activity involving the growth of seedlings could be presented to pupils in three different ways:

A. Place a set of seedlings into a dark cupboard, place another set in the light. Control all the other factors; see which produces the greenest shoots after a week.
B. Find out whether light and temperature affect plant growth.
C. Find out the factors that affect plant growth.

In all three cases the context (topic or area of interest) has been defined by the teacher; pupils have not been given the freedom to identify the topic as one on which they want to work. However, in A and B the problem is also defined for pupils by identifying the variables involved; not so in C which is, therefore, more open. A is more closed than B because the methods are defined, as is the dependent variable (colour of shoots). The number of acceptable solutions to the problem increases from A to C; clearly in C pupils may choose to investigate light, temperature, soil moisture, mineral nutrients, etc.

Open-endedness is achieved in an investigation by the teacher allowing pupils to make decisions, a process which could be seen as one of relinquishing control of the situation.[11] Is this really what is being suggested? Can the pupils cope or will there be chaos? Answers to these questions are 'yes', 'yes' and 'not necessarily'! Pupils will have experience of investigating from the age of five, so it is important not to adopt the traditional secondary school attitude of assuming that they know nothing when they come to you from their primary schools! We are looking towards incorporating open-ended investigations into the work on some occasions, not every lesson. Investigations will be part of the normal science programme rather than 'bolt-on' extras. That means that they grow naturally out of topics in the scheme of work that will be the major source of ideas for potential areas of investigation.[12] Relevant knowledge, understanding and process skills are prerequisites for carrying out investigations. In the previous example, tackling the investigation presented as C would be appropriate for pupils who have covered work on plant nutrition in Sc2 and who have developed skills of fair-testing. In this particular example, it is likely that the whole class, working in groups, will tackle the investigation more or less at the same time. Other situations will arise when a number of different investigations will be identified from an activity that initially has been controlled by the teacher. Examination of a bag of leaf litter that has been brought into the laboratory by the teacher might be used to address the initial question 'What animals are found in leaf litter?' Pupils would hand-sort this material, collecting the large numbers of invertebrates to be found, followed by identification of some of them using pictures and keys of various types. Pupils will raise a huge variety of questions about this material, many of which could form the basis of investigations. Some examples of the questions raised by year 7 and 8 pupils are:

What do woodlice eat? (Or snails, slugs, worms, spiders.)
Are different 'sorts' of woodlice found in different places in the wood?
Do they prefer some conditions to others?
Does the number of – vary during the year?
What happens to the leaves in the litter?
Do different types of leaves rot at different rates?
Will seeds germinate in leaf litter?
Do different seeds germinate at different times?

Pupils could be encouraged to produce a class list of these questions from which they might select their own investigations to work on for a period of time specified by the teacher, perhaps two or three double lessons.

Although the working methods will differ in different investigations it has been possible to identify a framework that is useful for structuring and planning investigations. The OPENS project suggest that there are five key stages involved in this process:

1. *Focusing*. The teacher sets up the situation that will encourage the class to ask questions and to identify the aspects of the question which they wish to investigate.
2. *Exploration*. Pupils structure and carry out their investigations after a process of clarification and speculation about issues such as fair-testing, testing predictions, etc.
3. *Reporting*. Pupils review their findings with the class, clarifying the findings and nature of the investigation.
4. *Consolidation*. Teacher and pupils use the information gained to develop knowledge and understanding and skills and processes, perhaps reflecting on any changes in views held.
5. *Application*. Identification and perhaps conducting of further investigations using the newly developed ideas and information.[13]

Some further issues involved in organizing investigations

Productive work

Pupils will not always raise questions that can be answered within the constraints of time and the resources at their disposal. They may suggest that they would like to investigate something which the teacher knows is impracticable, which will take too long, is unsafe, illegal, or cruel. In all these instances guidance by the teacher is vital. Unproductive questions are likely to be raised in situations like the leaf litter example above. However, so many possible investigations will have been suggested that pupils can easily be diverted on to a more feasible area of work.

Group size

Most practical activities in science involve pupils working in groups of two or three. Usually these groups are self-selected, a situation that most teachers adopt unless there is a specific reason for structuring the groups. Self-selected groups of this sort, which are usually based on friendship patterns, tend not to include pupils of widely disparate abilities. However some further consideration of group size is useful when investigations are involved.

The teacher needs to get round the groups to discuss their ideas as they get to grips with planning their investigations. If the focusing stage of the activity has been handled appropriately, the pupils will have begun their own exploration of the ideas. Nevertheless these will need to be reviewed by the teacher, who may well use such discussion as a checking point before pupils are allowed to start handling any apparatus or materials. When a large number of groups are involved, it may be difficult to engage with each of them at this crucial stage of the investigation. Slightly larger groups of five or six pupils might make this more feasible. Such groups can split up into smaller ones later if necessary. Another factor that has a bearing on group size is the complexity of the group's investigation. There is clearly an appropriate match to be found between the number of tasks involved and the number of pupils available to perform them. All pupils should be gainfully employed, jobs should be shared and rotated in different investigations or else someone becomes the permanent scribe, 'go-fer' or even onlooker.

Giving guidance and support

1. Equipment

This does not mean telling them what to do! In more closed practical situations pupils are usually briefed and given a list of apparatus and materials that they need for the exercise. It may well take some time for them to get used to a situation in which they are invited to use whatever equipment is in the laboratory or to produce their own apparatus list for the next lesson. Often they are able to take cues from the materials with which they are invited to work; hence their planning of the investigation is aided by interaction with some or all of the equipment that may be used. Problem-solving is an iterative process: once one finds that the first attempt fails, one gains a clue about a second possible solution.

2. Planning sheets

Conventionally, writing up of practical activities in school science has involved the use of a series of headings such as 'method, results, conclusions' or for younger pupils 'what we did' and 'what we found out'. Planning sheets of a more structured nature are useful in encouraging pupils to focus on the key areas of their investigations. They also function to remind pupils about the need to conduct fair tests, etc. so providing prompts about the nature of the processes of investigation. A wide range of such planning sheets are in use,[14] one example of which is shown in Fig. 3.2 (see pages 36 and 37). It is essential that the purpose and method of use of such sheets is explained to pupils. And remember, they will not complete them perfectly the first time that they use them. It is also important that such sheets are user friendly, so questions should not be asked on the sheets that pupils cannot answer. This may result in the use of a series of such sheets that become more sophisticated (and address higher level SoAs) as the pupils' experience of performing investigations and using the sheets develops.

Hence asking them to say 'why you think this will happen using scientific ideas' is appropriate to put on the planning sheet of a pupil who you reasonably expect to be working at or above level 5, but not on the sheet of someone at level 3.

Catering for different abilities

A number of factors determine the level of difficulty of an investigation apart from the extent of the decision-making about it that is demanded from the pupils.

1. The level of the concepts that are drawn on to plan, carry out and interpret the results. There is a broad parity between the levels at which pupils are working in the content areas of the National Curriculum and the levels in Sc1; hence the concepts that have been mastered by the pupils are needed to conduct investigations. For example, an investigation that is at level 5 involves pupils in having to 'formulate hypotheses where the causal link is based on scientific knowledge, understanding or theory'. Without having that knowledge, etc. they will be unable to tackle an investigation at level 5.
2. The nature and number of the variables involved. Independent variables such as light/dark, which are categoric in nature, are the basis of investigations that are easier to carry out than when the variables are continuous. This is used in the National Curriculum to differentiate between levels 5 and 6 in Sc1 strand (i). Hence an investigation in which plant growth is compared using batches of

UNIVERSITY OF HULL
SCIENCE EDUCATION CENTRE
PLANNING SHEET

QUESTION RAISING / HYPOTHESISING
Our question / hypothesis

What do you think will happen?
Why do you think it will happen?

PLANNING AND INVESTIGATING What are we going to do:	
These things will vary:	These things will always be kept the same:

OBSERVATION AND MANAGEMENT This is what we are going to measure / observe:

RECORDING How are we going to record our observations:

Figure 3.2 Planning sheet for science investigations

RECORDING What we observed / our results:

INTERPRETING What answer did your investigation lead you to?
How would you improve your investigation if you could do it again?
What would you like to do now to extend your investigation?

Figure 3.2 (continued)

seedlings kept in full light and dark conditions has a ceiling of difficulty at level 5. The more variables that have to be taken into account in an investigation, the greater its difficulty. This is used as a criterion for level 7 investigations, where pupils have to 'manipulate or take account of the relative effect of two or more independent variables'.

3. The complexity of the results. The nature of the data obtained in the investigation determines the difficulty of the task. Where conclusions are drawn that 'link patterns in observations or results to the original question', the level is 4. Where results allow reference to a scientific model, that part of the investigation accesses level 6.

From this, you will be able to see that any particular investigation that you wish to use can be pitched at a number of different levels using these factors. If you are dealing with a class in which there is a widespread of potential attainment, a 'mixed-ability' one, it will be necessary to ensure that all are working at appropriate levels of difficulty. Achieving such a match between curriculum experiences and pupils' ability involves stretching them to an appropriate degree; it does not involve giving them work that makes few demands on them. Two strategies for handling this situation present themselves. Investigations could be structured to different levels of difficulty by the teacher in advance of the teaching. Based on a prior knowledge of the class, different versions of the task are then given to different pupils. This process of differentiating the work by task clearly depends on the teacher having an intimate knowledge of the capabilities of each member of the class. The danger here is of long-term labelling of the pupil who subsequently works within these prespecified tramlines of attainment. Alternatively, a common investigation can be explored with the class. The demands of the task are then differentially met by different pupils or groups of pupils who find their own level within the activity. The great advantage of this approach is that pupils' attainment has not been predetermined by the teacher. Care must be taken, however, that an inappropriately low ceiling of attainment is not imposed by the nature of the investigation set. If the investigation seems to be too difficult, pupils can be cued by the teacher either through prompt cards or through oral interactions.

What is essential in managing the conduct of investigations is to be aware of the possibilities that exist for extending the investigations, coming up with ideas for new ones or using them to revisit areas of difficulty previously identified.

Notes and references

1. Friedler, Y. and Tamir, P. (1990) Life in science laboratory classrooms at

secondary level, in Hegarty-Hazel, E. (ed.) *The Student Laboratory and the Science Curriculum*. London: Routledge.

2. Hodson, D. (1992) Redefining and reorienting practical work in school science. *School Science Review*, vol. 73. no. 264, 65–78.
3. Driver, R. (1983) *The Pupil as Scientist?*, p. 9. Milton Keynes: Open University Press.
4. Dallas, D. (1980) *Teaching Biology Today*, p.68. London: Hutchinson.
5. Schools Examination and Assessment Council (1992) *Science 1992 National Pilot, School Assessment Folder (Part 2)*. London: SEAC.
6. Woolnough, B. and Allsop, T. (1985) *Practical Work in Science*. Cambridge: Cambridge University Press.
7. Perrott, E. (1982) *Effective Teaching: A Practical Guide to Improving your Teaching*. Harlow: Longman.
8. Dip a privet leaf into a small amount of propanone in a beaker. Immediately place the leaf on a square of overhead projector acetate (try different sorts). Exert downward pressure for 20 seconds using a rubber bung. Remove the leaf and examine the imprint.
9. Many sources are available; particularly useful is Troughton, J. and Donaldson, L.A. (1972) *Probing Plant Structure*. London: Chapman & Hall.
10. Jones, A.T., Simon, S.A., Black, P.J., Fairbrother, R.W. and Watson, J.R. (1992) *Open Work in Science: Development of Investigations in Schools*. Hatfield: Association for Science Education.
11. Lock, R. (1990) Open-ended, problem-solving investigations. What do we mean and how can we use them? *School Science Review*, vol. 71 no. 256, 63–72.
12. Sources of ideas can also be found in many published texts, e.g. Bedfordshire Activities for Process Science 1, 2, 3, Bedfordshire Education Service, 1990. Rowlands, D. (1987) *Problem-solving in Science and Technology*. London: Hutchinson.
13. OPENS project, Jones, A.T. *et al.* (1992) op. cit.
14. Other examples are in Jones *et al.* ibid., and SEAC (1992) *School Assessment Folder (Part 3), Materials to support the assessment of Sc1: Scientific investigation*.

CHAPTER 4

A variety of teaching approaches

Our PGCE students often tell us that one of their greatest areas of difficulty is in teaching those topics where there is no obvious practical work to draw on. It is difficult, then, to make the learning an *active* process for the pupils. In these circumstances it is easy to resort to activities such as dictation of notes to the pupils, copying from blackboard or textbook or completing an undemanding worksheet. This happens particularly when dealing with factual information that the teacher feels must be transmitted to the pupils, usually for examination syllabus coverage requirements. Science teachers claim themselves to be heavily committed to practical work, which is seen as the fundamental basis of learning science, yet the proportion of lesson time devoted to practical activities is surprisingly low. Table 4.1 shows the proportion of time allocated to different activities found in the 'Effective Use of Reading' project.[1]

Table 4.1 ***Time spent in different activities by seventh and tenth year pupils as percentage of total science***

	Year	
Activity	7	10
Practicals	23	11
Observing	7	8
Reading	9	10
Writing	11	20
Listening	26	31
Non-involved	8	10
Discussion	13	10

Activities related to processing of information (reading, writing and listening), rather than to practical experiences occupied a total of 46 per cent of pupil time in year 7 and 61 per cent in year 10. Sutton[2] has pointed out that practical work, and what he calls 'word work', should both be directed towards the process of extracting ideas – an interpretive activity that engages the mind of the learner. Similarly, Klopfer has warned 'of making sharp distinctions between learning outcomes to be expected from the science laboratory and outcomes from non-laboratory classroom work'.[3] Whenever possible pupils should be challenged by tasks that place them in active rather than passive modes of response: having to *do* something as well as listen.

Teaching through the use of 'second-hand' data

There will always be a place in school for the textbook. Textbooks traditionally present a map of an area of knowledge pitched at a level of complexity that, hopefully, is appropriate for the reader (both in terms of conceptual difficulty and ease of reading[4]). That is, they make clear what is known about a subject, e.g. the functions of the kidney are nitrogenous excretion, osmoregulation, etc., or blood is made up of two main types of cells.

Thirty years ago, Joseph J Schwab, the supervisor of the American Biological Sciences Curriculum Study (BSCS) Handbook[5] described this kind of treatment as 'the rhetoric of conclusions' and pointed to some of its limitations:

> Both by omission and by commission it gives a false and misleading picture of the nature of science.
>
> By commission, a rhetoric of conclusions has two unfortunate effects on the student. First, it gives the impression that science consists of unalterable, fixed truths. Yet this is not the case. The accelerated pace of research in recent years has made it abundantly clear that scientific knowledge is revisionary. It is a temporary codex, continuously restructured as new data are related to old.
>
> A rhetoric of conclusions also tends to convey the impression that science is complete. Hence, the fact that scientific investigation still goes on and at an ever-accelerated pace is left unaccounted for to the student.
>
> The sin of omission by a rhetoric of conclusions can be stated thus: It fails to show that scientific knowledge is more than a simple report of things observed, that it is a body of knowledge forged slowly and tentatively from raw materials. It does not show that these raw materials, data, spring from planned observations and experiments. It does not show that plans for experiment and observation arise from problems posed, and that these problems, in turn, arise from concepts which summarize our earlier knowledge.

A major innovative dimension of BSCS materials, which was followed in Britain in the Nuffield O and A level projects, was the use of a 'narrative of enquiry'. Schwab characterized this as follows:

> That is, our current views on a subject such as genetics are developed step by step through a description of the experiments performed, the data obtained, and the interpretations made of them (Schwab 1963, p.40).

Table 4.2 **Summary of a BSCS Invitation to enquiry: control of blood sugar**

Teacher input	*Students' task*
1. Revision of ideas about the role of glucose in the body; its origin in food; digestion and absorption in the gut. Describes experiment in which laboratory animals are all fed and exercised, during which blood sugar level is monitored →	Draw a graph to predict blood sugar levels based on the information supplied
2. Clarification of issues raised. Presents further experimental results relating to the glucose 'balance sheet', which indicate differences between expected and actual blood sugar levels →	Speculate about the discrepancy, raise possible reasons for it
3. Presents details of blood circulation between gut and liver, in the light of which alternative reasons, given by students, are evaluated leading to identification of a hypothesis about the role of the liver →	Decide how to test the hypothesis, what data would need to be collected and what patterns in the data would support or refute the hypothesis
4. Gives the results of such an experiment, i.e. Gut → (+) Liver → (–) blood glucose level Presents alternative possibilities of the role of the liver in storing or destroying glucose →	Identify data which would support the storage hypothesis
Presents details of the glucose/glycogen balance and biochemical detail of the structure of glycogen →	Evaluate the extent to which the storage function is supported

BSCS published *Invitations to Enquiry*, which placed students in situations by which they came to appreciate not only *what* is known but also *how* it is. Latterly Sutton has argued for materials of this sort, resting on the interpretation of ideas and their significance, to receive special emphasis, because they are becoming 'an endangered species in need of special protection and cultivation'!

Two examples of this approach follow. The first (Table 4.2) is a summary of a BSCS invitation to enquiry, written originally for sixteen to eighteen year old students, involving the control of blood sugar in mammals.[6]

The 'invitation' goes on in a similar way to probe the role of skeletal muscles and fat and protein conversion into glucose. Two further invitations to enquiry deal with the mechanism of homeostasis involving blood sugar. The factual details of the functions of the various organs are drawn out from the narrative of enquiry, but the students are obviously learning much more than that. Clearly an

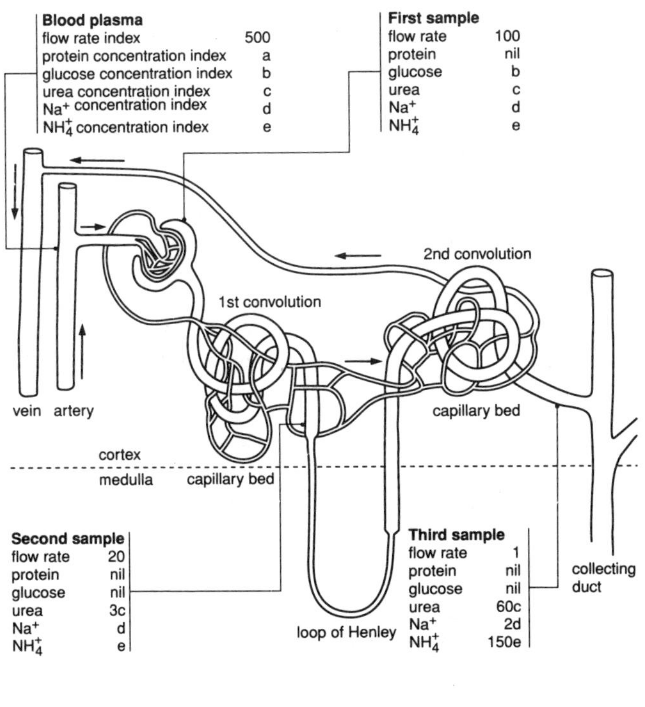

Figure 4.1 A simplified diagram of a nephron and part of its blood supply. The samples show what is happening at various points in the tubule.

alternative expository approach could be adopted by the teacher by stating (or listing on the blackboard, in a handout, by dictated note or copying from a textbook) the functions of the liver, etc. For the students involved, the first approach is likely to engage them in a way that the second one will not.

The second example appeared originally in the Nuffield O-level Biology course[7] and has been used and adapted extensively since then. Despite heavy conceptual demands, with skilful preparation and handling by the teacher, the approach is appropriate for fourteen- to sixteen-year-olds. Following work on the gross and microscopic structure of the mammalian kidney, the function and mechanisms involved are tackled by inviting the pupils to take part in a 'second-hand' investigation. They are given information about the total volume of filtrate produced in the kidneys, the volume output of urine per unit time and the microtechniques by which samples of filtrate can be withdrawn from different regions of individual nephrons. They are then presented with the information shown in Fig. 4.1 and required to compare samples taken successively along the nephron.

Before they start pupils must understand (i) that flow rate index is a measure of how much fluid passes a particular point in a given time (not too difficult, various analogies are obviously possible) and (ii) the concept of concentration.[8] Their comparisons then enable them to work out the pattern of reabsorption of water, glucose, etc. and hence the functions of the various regions of the nephron. They perceive the significance of the retention in the blood plasma of its proteins and understand that the constitution of the filtrate can be changed by reabsorption of water and solutes and by secretion into the lumen of the nephron. Again these knowledge outcomes are gained by the active processes of thinking (puzzling) and through discussion; ways in short of making the learning meaningful to the pupils. Many other examples of this approach can be found in the newer school textbooks; they can usually be recognized by the fact that they ask genuine questions of the reader, hence bringing him or her into a dialogue with the writer. Good material of this sort needs to be collected for your resource bank of teaching ideas.

Pupils' writing

Every year when talking with PGCE students about pupils' written work it is obvious that the assumption is being made that the main sort of written work for pupils is the experimental account. Someone soon advances the view that a standardized format should be used for this, with subheadings such as 'method, results, conclusions'. The past tense is required, as is the passive voice and personal pronouns should be avoided. When these issues are questioned, a heated discussion inevitably follows in which the clinching justification is given that 'That's how scientists write research papers' and therefore that is what youngsters should be trained to do!

In a much-quoted article, Medawar has suggested that the scientific paper gives a stylized and somewhat misrepresented view of science anyway, focusing as it does on the justification of the conclusions drawn from the research data. This gives the impression, as Sutton says, that the 'thinking comes mainly after the looking and doing'.[9] If one of the objectives of school science is to introduce pupils to the methods of science then clearly it is important that misrepresentation of those methods is avoided.

The purpose of a scientific paper is to communicate the findings of research to the academic community of which the writer is part. What then are the purposes of pupils' writing? Teachers have given the following reasons for setting writing tasks:[10]

> Writing helps children think logically through an argument.
> As a record of work done.
> As a means of communication to others.
> Writing notes for later revision.
> Writing helps pupils to learn facts and concepts.
> Extra thought is involved in writing over and above that for talking.
> Writing can be used to encourage observation.
> Writing tells the teacher when pupils have not understood something.
> A chance for pupils to extend, speculate and produce new ideas.
> A means of assessment.
> Practice for answering examination questions.
> Writing can encourage reading.
> To encourage discussion.

Most of the reasons given are related to the enhancement of learning, which can come from setting *appropriate* writing tasks. A Schools Council Project, *The Development of Writing Abilities 11–18*, has analysed a wide range of written material collected from secondary schools, in terms of the function of the writing and its sense of audience (who was the intended reader). One sort of writing, *transactional*, has the function of communicating the outcome of tasks such as recording, reporting, classifying information. Young pupils find this sort of writing which includes the traditional experimental account, difficult to handle particularly because of its cold formality. *Expressive* writing, on the other hand, is much more personal, closer to the language of talk and includes the writer's thoughts and feelings. The message is clear:

If, as seems likely, work needs to be developed and communicated by language which is appropriate to the character of the work, then the observing, thinking and investigating in science should initially be articulated not through transactional but through expressive writing, which, like the work, is both personal and exploratory.[11]

Such writing is particularly suited to problem-solving, investigative practical activities, in the early stages of which ideas are raised and discussed and possible procedures are clarified. This is when 'thinking with a pencil in hand' occurs, a process accompanied by changes of tack, jotting, revision, etc. To impose a formal style of writing on pupils at this point is likely to be counter-productive; the planning sheets or other methods used for writing up come later after a process of reflection and redrafting. Even then, the writing does not have to be transactional.

As pupils move through the secondary school there is a need for them to become familiar with and competent in transactional writing. They should recognize the form of it (as that used in their textbooks) and come to see the purpose of it (for example, in communicating essential facts or findings as in a newspaper report or a scientific paper). They should also acquire the skills involved in producing it. In this they will need structured support; the teacher cannot assume that they will be able to write in this way because the style is obvious to him or her. Strategies such as identifying and listing key points of the method and results, having pupils sequence them using flow charts and other forms of diagrams, etc. are essential procedures for developing these skills.

Note-making or note-taking

Having a good set of notes for test or examination revision purposes appeared in our list of reasons given by teachers for setting writing tasks. In the past far too much time has been spent by pupils on copying from books and by taking down teachers' dictation.[12] Both activities are essentially passive and some pupils like them, though they lull pupils into believing they are working hard. The dictated word leaves the teacher's mouth and appears on the pupil's page by a process that utilizes hand and pen but does not engage the brain! Lesley Bulman makes the point with great clarity:

Copying or taking dictated notes not only inhibits pupils' understanding of science and their ability to communicate, but also may be one of the reasons why pupils find school science so dull and difficult and they give it up at the earliest possible opportunity.[13]

Is there *any* place for either of these activities? Perhaps there is when it is essential that the pupils master a demanding and precise definition or term. The definition of osmosis is such an example, which could be dictated, or copied from chalkboard, overhead projector or textbook. Opportunity will exist after this for the pupils to express examples of the phenomenon in their own words.

The process of note-making is usually preceded by or connected in some way with that of reading. Reading also can be either an active process or a passive one. Active reading is likely to occur when the reading is done for a specific purpose, usually the subsequent completion of a writing task. Passive reading is encouraged by an instruction such as 'make your own notes from pages X to Y in the textbook'. This encourages wholesale copying of chunks of text. Active reading, on the other hand is prefaced by instructions such as:

Find and mark all references to . . .
Underline in red all the words which . . .
Label on the diagram, all the parts which . . .

Such task-orientated approaches or *directed activities related to text* (DART), encourage active, reflective reading. They can lead to means of recording of information by pupils that fulfil many of the purposes listed earlier. An example of a DART activity is given below and was intended for older students. It draws on some seventeenth-century writing and therefore meets a National Curriculum requirement to visit important ideas in the history of science (PoS, Key stages 3 and 4).

Read the following passage (taken from *Of the motion, action, and office of the heart* (William Harvey, 1628)) and then do the tasks listed under it.

First of all, the auricle contracts, and in the course of its contraction throws the blood (which it contains in ample quantity as the head of the veins, the storehouse and cistern of the blood) into the ventricle, which being filled, the heart raises itself straightway, makes all its fibres tense, contracts the ventricles, and performs a beat, by which beat it immediately sends the blood supplied to it by the auricle into the arteries; the right ventricle sending its charge into the lungs by the vessel which is called vena arteriosa, but which, in structure and function, and all things else, is an artery; the left ventricle sending its charge into the aorta, and through this by the arteries to the body at large.

These two motions, one of the ventricles, another of the auricles, take place consecutively, but in such a manner that there is a kind of harmony or rhythm preserved between them, the two concurring in such wise that but one motion is apparent, especially in the warmer blooded animals, in which the movements in question are rapid.

1. Underline in red the parts of the heart that carry out an action.
2. List these in the order in which they are named in the passage on the left side of a page.
3. Underline in blue all the parts of the heart that receive blood.
4. List these in order opposite the parts in your first list.
5. Join up the two lists with arrows to show the direction in which blood flows.

Table 4.3 **Directed activities related to text**

Reconstruction activities (using text modified by teacher)	*Analysis activities (using straight text)*
Pupil task: pupils complete text or diagram, reconstructing meaning	Pupil task: Pupils locate and categorize text information by marking and labelling. Use marked text as basis for summary (diagrammatic or note form)
1. Text completion (a) Word completion (selected words deleted from text) (b) Phrase completion (selected phrases/clauses deleted from text) (c) Sentence completion (selected sentences deleted from text)	1. Text marking Locating and underlining parts of text representing certain meaning of information targets
2. Sequencing (a) Scrambled segments of text arranged in logical/time sequence (text cut into segments representing steps/events, etc.) (b) Segments of text classified (texts cut into segments representing certain categories of information)	2. Labelling Pupils label parts of text using labels provided by teacher 3. Segmenting Pupils break text into meaning or information units and label/annotate segments of text
3. Prediction (a) Pupils predict next event/step or stage after reading segments of text (text segments presented a section at a time) (b) Pupils write next part or end of text (text presented a section at a time)	4. Table construction Pupils produce column and row headings for tables and fill in cells using text(s) as source of information 5. Diagram construction Pupils construct and complete diagram appropriate for particular text, e.g. flow diagram for text describing a process, branching tree for a text describing a hierarchical classification, networks, etc.
4. Table completion (a) Pupils fill in cells of table using row and column headings and text as sources of information. (Teacher provides row and column headings) (b) Pupils devise row and column headings using texts and cells of matrix as source of information. (Teacher fills in cells)	6. Pupil-generated questions Pupils read text and generate questions they still need answers to 7. Summary Pupils produce headings and summarize information
5 Diagram completion (a) Label completion using text and diagram as sources of information (selected labels deleted from diagrams) (b) Diagram completion using text and partly complete diagram as sources of information. (Teacher constructs original diagram: flow diagram, branching tree, network, etc.)	

Source: Davies F. and Greene T. (1984) *Reading for Learning in the Sciences.* Oliver and Boyd.

6. Add the names of the blood vessels that carry blood away from the heart.
7. Use your textbook to find out the modern name for Harvey's 'vena arteriosa'.
8. Find out where blood goes to when it leaves (i) the lungs and (ii) different organs in the body such as kidneys.
9. Harvey said that the action of the auricles and ventricles takes place 'consecutively'. What does this mean?
10. How would you describe the action of the two ventricles?

The important book by Davies and Greene[14] lists and gives examples of the range of possible DART exercises, which appear in Table 4.3. They also suggest the following practical ways by which pupils may mark text:

1. direct marking of photocopied pages (permission to photocopy having been obtained from the publisher);
2. by inserting pages of text into acetate covers on which marking can be done using water-soluble pens;
3. using overhead projector acetates to overlay text.

Method (1) has the advantage of forming a permanent record in the pupils' files.

Writing for different audiences

The Schools Council Writing Project, in their research published in 1976, found the use of transactional writing greatly to exceed that of expressive writing across all subjects of the curriculum; the incidence of the latter in the science lessons observed could hardly have been lower, because it was nil! Things have moved on since then. However, a further feature of the writing in science was that it was presented as if written for an intended audience of the teacher in the role of examiner. This involves pupils telling the teacher what the teacher knows already. Watkins points out that the pupil knows that the teacher knows and hence the purpose of the writing is to enable the teacher to *assess* what is said. In that sense it is not a piece of genuine communication and it is unlikely to produce motivation.[15] Many writers advise teachers to try to incorporate styles of writing into pupils' written work, where writing is done for audiences other than the teacher. Other potential audiences are the pupils themselves, younger or older pupils (pen-friends, classroom peers), adults, e.g. newspaper editors, managing directors of industries, drug companies, etc. Written work might include activities such as writing poems, keeping diaries, writing newspaper reports, stories, letters, scripts for plays, TV commercials and so on. There are many examples available of highly imaginative responses to exercises of these sorts.[16] However, if such writing tasks are going to be of value to the pupil they need to be both enjoyable to do and to offer opportunities for learning to take place. There is a danger that such exercises, if confined to the odd homework or the last lesson before the Christmas holidays, will be seen as trivial by teachers and pupils alike. Such activities are more successful when the writer's sense of audience is clear; the activity has been prepared for appropriately (thought through and talked over) and the teacher responds supportively to the product.[17] This is most likely to happen when the audience is a real one rather than imaginary and the communication is genuine. With these requirements in mind the most readily available non-teacher audiences are likely to be the pupil and his or her peers. For the individual pupil the learning benefits that are derived are likely to be the greater understanding of concepts that comes from synthesizing them in order to meet the different requirements of such tasks. Writing of this kind still allows the teacher to identify how much learning has taken place and the nature of any misconceptions held. The extract shown below is by a year 10 pupil responding to the invitation to do some imaginative writing based on the well-used theme of miniaturizing people and, in this case, travelling through the blood system. Try analysing the extract to pick out all uses of correctly learned concepts.

A journey around the circulatory system

To explain the meanderings of a red blood corpuscle on its long journey through the circulatory system I thought it would be picturesque to imagine that I am sitting on one.

I find myself in a huge chamber with millions of other red blood cells. The noise is terrific, gurgling, whirling and there is a roaring in my ears, a huge leathery trap door lies ahead, the pressure on my ears seems to grow.

Suddenly the trapdoor opens and I find myself flying into a second chamber. If I thought the pressure was high in the first chamber it seems much higher here. Muscular walls close in towards us. When I think I can stand it no longer another trap door opens and millions of us zoom out along a giant causeway. Soon this main causeway begins to branch into many smaller passageways, some of my companions fly off in one direction, some in another. The tubes become smaller and smaller until I could reach and touch the roof with my hand. Now there is only room for one red cell at a time. Through the near transparent walls I can see colourless fluid oozing out, this must be the spongy lung that I have read about. After what seems an age the tubes grow larger again, there is room for several of us to roll along together. But my companions look different, they are bright red in

continued

A journey around the circulatory system *continued*

colour, I'm sure they were dark red before. Now I find that fluid seems to be seeping towards us through the walls instead of away from us. The pace quickens, the passage becomes bigger and bigger. Now we are joined by millions of red blood cells.

Then! Oh dear, we are back into another chamber. We go through the same process that I suffered not less than a few seconds before, through a trap door into a second chamber and then whoosh, spinning along through a giant pipe. We whiz onwards many passages bearing off towards the huge pulsating intestine. But I carry on with millions of other companions, the large majority in fact, into the kidneys. Again the passageways become smaller and smaller as I experienced before, fluid out through the walls and then fluid back in again; a big division of the ways and down into the right leg. This seems to be the longest journey yet, same old story, passages growing smaller and smaller until there is only room for me, then slightly bigger again. But! what on earth is lying ahead. There are hundreds of cells shaped a bit like the one I'm on but they are white in colour and they seem to be attacking something, no business of mine however so I just speed on.

Good heavens! now we have a snag, the way ahead is blocked, or nearly so. Thank goodness we can squeeze past. We struggle past a huge long jelly like clot, like a snake taking up most of the room in the passage way. If this breaks off and speeds along with us towards the heart my host is going to be in trouble.

Am I seeing things or have we all changed colour again, we all seem much darker than we were before, we must be going up to the heart again which will push us on for more oxygen in the lungs.

Learning by talking

Here is a piece of classroom talk between two year 8 pupils doing some practical work:

Pupil 1: 'Right, what do we do next?'
Pupil 2: 'Test it with iodine.'
Pupil 1: 'Where's the iodine? Has it gone blue black?'
Pupil 2: 'I don't think so, what do you think?'
Pupil 1: 'No, there's no starch there.'
Pupil 2: 'OK . . .' (starts to write down the results).

The pupils, through their talk, are helping each other with the sequence of a practical procedure and clarifying their observations and conclusions to their mutual benefit. Talking like this is obviously central to pupils' learning and should be encouraged. It leads to the pooling of ideas, sometimes to productive argument over disagreements and allows previous knowledge to be brought into play in a new situation.[18] It has important *social* aspects, too, in contributing to the production of amicable working relationships, the equivalent of grooming in other primates! Such talk is not deliberately set up by the teacher, it is an inevitable complement of small-group practical work. The advantages gained from such discussion can also be drawn from situations where no practical work is involved and where the discussion is set up by the teacher, as in the following examples.

Brainstorming

This is a useful device that usually serves as an introduction to a lesson or a topic. Its function is to generate a number of ideas, for example what is known already by the group or what might be done in a problem-solving situation. The ideas should be recorded on paper, overhead projector transparency or flip-chart as they emerge. Since the purpose of the activity is to generate ideas for later discussion and follow-up, ideas should be accepted from all without further comment or clarification. Group size should be six or seven and a time limit of five minutes or so should be specified for the activity.

Discussing

I have in mind here situations where small groups of pupils discuss a topic that has been identified by the teacher or by the pupils themselves. This *excludes* teacher interaction with the whole class. For such discussions to be productive the topic must be one that is motivating and relevant to the pupils and in which genuinely different views can be held and explored. One note of caution: if pupils are not used to discussion in class, pitchforking them into talking about issues such as genetic engineering or the use of condoms is likely to be disastrous! Start with something that is nearer to their normal pattern of working such as the different interpretations which can be given to pupils' experimental results.

Social psychologists tell us that interaction between individuals in small groups continues as long as the exchanges are seen by all participants as being profitable. For this to happen the contribution made by each individual must be seen to be valued rather than derided. The implication of this is that discussions should be managed by the teacher and not simply left to happen. Just as the teacher has an important role in allowing openness to occur in practical activities such as investigations by avoiding over-structuring the situation, so the same holds true in handling classroom discussions.[19]

Researchers in the Ford Teaching Project[20] made the important point that teachers need to adopt two different sets of classroom strategies to encourage genuine inquiry and free discussion:

constraint-removal and *guidance strategies*, i.e. what *not* to do as well as the need to make positive interventions.

Teachers may impose constraints during discussions by:

1. calling upon pupils by name and hence forcing a reluctant contribution;
2. setting up question and answer sequences and closing down the discussion;
3. imposing consensus by utterances such as 'Do you all agree with that?' or by rewards such as 'yes', 'right', 'good';
4. asking supplementary questions that seek clarification from the pupil and which may be perceived as adverse criticism;
5. breaking silences too early and forcing the pace of the discussion;
6. allowing pupils to make personal comments on others' contributions;
7. over-suppressing dominant pupils.

Peer group constraints may also operate and the teacher needs to be alert to the existence of possible gender expectations. These may surface in a variety of different ways. Boys may regard it as being inappropriate for girls to have a view about certain subjects and vice versa. Indeed, both sexes may feel that they are not prepared to participate in discussions where the topic is felt to be gender-specific. The timing, frequency and quality of contributions may also be perceived as being inappropriate, particularly by boys of those of girls, and considerable pressure may be brought to bear to ensure that such contributions are curtailed.[21] Sensitive handling of these issues is demanded of the teacher who will have to make a judgement about whether to tackle them by making the existence of such pressures explicit to the groups and hence allow them to become a topic for group discussion in itself. Alternatively, the teacher may operate by having a quiet word in the appropriate ear at the right time.

Practical considerations such as group size, membership and seating arrangements are also important for groups to function effectively. In most circumstances, and unless there is a good reason for doing otherwise (e.g. control of difficult pupils), groups will be self-selected and hence based on friendship patterns. Size of group will depend to some extent on the nature of the topic. In general, the smaller the group the greater is the freedom for discussion but also the more restricted the range of ideas is likely to be. Opinions vary about group size – for most purposes three to five seems to be appropriate but some trial and error on your part is demanded here. What is important is that they can all sit comfortably and see and hear each other. Room geography may pose problems and these will have to be circumvented appropriately.[22]

Topics for discussion

Earlier it was said that discussions should be based on topics that are motivating because of their relevance to the pupils and in which different views can be held and explored. Hence topics that lend themselves particularly to this approach are science-based social issues. Increasing emphasis has been placed in all levels of education on science, technology and society over the last twenty years. Joan Solomon has reviewed this trend and identified the following features of the movement in science education:

An understanding of the environmental threats, including global ones, to the quality of life.
The economic and industrial aspects of technology.
Some understanding of the fallible nature of science.
Discussion of personal opinion and values, as well as democratic action.
A multi-cultural dimension.[23]

Some of the areas falling within these trends provide obvious and well-known topics for discussion, e.g. fluoridation of water supplies (and other aspects of community health), genetic counselling, organ transplantation, energy policies, etc.[24] Many of these topics involve sophisticated ideas and judgements of finely balanced opposing views. This being so, it is probable that discussion of areas of this sort will need to be prepared for by pupils working in advance on structured activities that provide them with an appropriate knowledge background. In the *Discussion of Issues in School Science* project the lead-in material for the discussions involved viewing a video. In this project, the pupils exercised control over the selection of the areas of the topic that they wished to discuss, which is unusual. This strategy seems to have been an effective one: on occasions the quality of discussion was described as 'vivid and memorable'.[25]

Whilst pupils may engage fully in discussions, and in that sense be involving themselves personally in the subject, can they be said to be learning anything from what can be a time-consuming type of activity? Teachers with reservations of this sort will see the need for some sort of follow-up exercise perhaps of a more formal nature, such as a short piece of writing that summarizes the main points at stake in which the science content is clarified. This may have the additional advantage of allowing the pupils to experience a tangible outcome from their discussion which has, then, an obvious point.

Drama and role play

The main difference between drama and role play is in the degree of scripting involved: in drama, detailed scripts are used and followed; in role play pupils work to briefs but not complete scripts. The range of possibilities for employing both as teaching strategies is very wide and an encouragingly large number of examples are available in the literature to draw on directly or indirectly.[26] The arguments for using such techniques are that pupils enjoy them and find them motivating[27] and they may well make sophisticated and controversial contexts more accessible to pupils and open up these areas for more effective discussion.[28] Despite the fact that pupils find these approaches motivating, they are not used extensively by science teachers, some of whom report them to be threatening because they lie outside their own experience and training.

In drama, pupils 'see the processes and consequences of a series of actions forming the drama as they are acted out in front of them'.[29] Pupils may or may not have been involved in writing the piece, but since the number of actors is likely to be small most pupils will be involved in watching the drama. Whether the actors are the pupils themselves or members of one of the occasional peripatetic companies that visit schools, what transforms the event into an educational experience is the involvement by all the pupils in discussing the issues raised by the drama.[30]

Role play probably has wider applicability in science teaching, which accounts for its greater use at present. By undertaking roles, pupils can experience something of the feelings and reactions of others in particular situations. Joan Solomon describes how a Scientific Eye television programme, 'Microbeasts and Disease', which includes a short cartoon sequence on the work of Edward Jenner, was used as the basis for a successful role play.[31] Role play exercises have been used profitably in decision-making simulations that have a biotechnology theme. Pupils' responses to teaching strategies of this type are generally favourable.[32]

For new teachers who are thinking of adopting teaching approaches of these sorts it might be useful to start from small beginnings. Try introducing short (e.g. five or ten minute) role play exercises into lessons on several different occasions so that pupils begin to get used to the idea. In other words do not abandon the approach if it appears not to work on the first occasion. Your colleagues who teach English may well be a good source of advice about role play and drama.

Games and simulations

These have become increasingly frequently used in school science over the last twenty years or so. The terms 'game' and 'simulation' are often used interchangeably but attempts have been made to categorize them more precisely. A game involves *competition* (between players or against the game's designer) that is directed by a set of *rules*, whereas a simulation represents a *real* situation involving some on-going process. An exhaustive review,[33] published in 1981, identified three ways in which science-based games and simulations might be used:

1. as aids to the teaching of the basic content of science courses;
2. for educating *through* science (using science-based exercises to cultivate useful skills and desirable attitudinal traits);
3. for teaching *about* science and technology and their importance to modern society.

These authors reviewed the range of materials available for use at that time. Apart from commercially available examples, the pages of *School Science Review* and *Journal of Biological Education* are sources of examples and ideas that have usually been tried out in the classroom. One essential aspect of the use of such resources is that the debriefing that must follow as an integral part of the activity should allow students to reflect on the relationships between the exercise and the real situation on which it is based. A particularly effective way of achieving this is by having pupils design their own games based on well-known examples such as Snakes and Ladders[34] or Trivial Pursuits. An impressive example of a board game devised by some GCSE pupils to exemplify the principles of food chains is described by Barlex and Carré.[35]

Teaching post–sixteen students

It is something of a paradox that the ablest students in biology, those who will go on to take university degrees in biological sciences and biology-related fields, have in the sixth form traditionally been exposed to some of the most unexciting of teaching approaches! A-level syllabuses, notoriously, are crammed with factual information that, with some justification, teachers feel that students must be able to regurgitate for examinations. Nevertheless, exactly the same arguments and issues about the need for active teaching approaches, which have formed the basis of this and the previous chapter, apply with equal force to sixth-form teaching.

Currently, something of a revolution is taking place in the post-compulsory curriculum. Vocational A levels in the shape of General National Vocational Qualifications (GNVQ) started in science in September 1993. New A and AS syllabuses are to be introduced in 1996 that will have the aim, among others, of bringing post-sixteen courses into reasonable end-on continuation from the National Curriculum.[36]

Two important initiatives have been underway for some time: these are the development of modular courses at A level and in flexible learning. Modular A-level programmes have become available for a number of subjects and, in science, have followed on from earlier popular GCSE syllabuses such as NEA Modular Science. They will be retained after 1996 when they will run alongside existing linear syllabuses at A and AS levels.[37] Flexible learning approaches have been incorporated into both modular and traditional post-sixteen courses and the rest of this chapter is given over to a brief discussion of them.

What is flexible learning?

Waterhouse[38] has pointed out that it is more important to identify the attributes of good teaching and learning within flexible learning initiatives than to try to define the proliferating terminology associated with them. He suggests that the following are such attributes:

1. *A variety of styles and techniques are regularly used.* This would therefore include inspiring whole-class teaching, as well as well-organized small group and individual work.
2. *Arrangements* that create the environment for teaching and learning are *flexible* with the needs of the student constantly at the fore.
3. The *resources* that provide the data and stimuli for learning should be drawn from as *wide a range* of sources as possible.
4. The *teacher* should create frequent opportunities to work *personally* with students in very small groups in order that his/her example and influence should be maximized.
5. In every way the work within classes should be *student-centred*, and flexibility should be used in order to achieve this.

This draws on a number of developments that have been made in the last twenty years in areas such as resource-based learning, independent learning, supported self study and open learning.[39] Currently there is a national project, the Flexible Learning Project, (supported by the Training Agency) that aims to promote flexible learning by working through a network of regional project leaders appointed by consortia of local authorities.[40]

The application and adoption of active learning strategies that employ the use of a wide range of different types of resources, independent learning and small group work clearly places the teacher in a different role to that of the traditional provider of information[41] and the student from that of a consumer of it. They do, however, require as effective classroom management as does any other teaching situation.[42] Denise Mountford[43] suggests that biology is a subject that lends itself to a flexible learning approach for the following reasons:

1. it is usually taught in a thematic way and can be broken up into short units of work within broader areas of study;
2. it is a subject that requires practical investigations and, therefore, individual and group investigations;
3. it is a subject for which there is a rich variety of resources in a variety of media;
4. it is a subject that raises many controversial issues that lend themselves to questioning and debate;
5. it is a subject for which the skills of research, independent study, problem-solving and communication are essential;
6. there are many environmental and technological aspects to biology which provide opportunities for cooperation with industry and government agencies.

At Wyke Sixth Form College in Hull, the Biology Department uses the approach in most areas of their teaching of the NEAB A-level syllabus. Teaching and learning involves different kinds of activities:

Practicals: in which students work individually or in small groups from briefing sheets that are laid out as follows:
- (a) introduction to the activity,
- (b) procedure to be followed,
- (c) discussion – structured by asking specific questions and extended by a short research task (either a practical follow-up or a literature activity).

Study items: independent or small group work that extends students' practical work largely through reference to sections of Nuffield A-level Biological Science Study Guides and other references.

Research tasks: individual work in which students are directed to particular resources in order to complete specific and structured tasks. These are longer than study items and are less closely related to practical activities.

Question items: examples of A-level questions on the areas of work concerned and completed in study time rather than class time.

Presentations: more formal whole group teaching that identifies and reinforces key ideas and issues, accompanied by an overview handout that itemizes the main issues to be considered and indicates what is to be accessed by the students.

Audiovisual or CAL tasks: resource-based tasks approached independently, by small groups or the whole group.

Tutorials: when consolidation and clarification of areas of the work occur.

How this works in practice is illustrated in the structure of the teaching of the topic 'Absorption and transport in plants'. The components are as follows:

Practicals

PL 1 Investigating the mineral requirements of plants
2 Studying the morphology and anatomy of root and shoot
3 Investigating the rate of water loss from a leafy shoot
4 Leaf structure and rate of water loss
5 Investigating the behaviour of stomata

Study items

S 1 Van Helmont's experiment on willow
2 Vegetable Staticks (Stephen Hales)
3 The rise of water in a plant

Research tasks

R 1 Mineral requirements in plants
2 Anatomy of leaves and environmental factors in water loss

Question items

Q 1 Structured question: mineral nutrition in plants
2 SQ: Anatomy of stem and root; movement in xylem
3 SQ: Transpiration
4 SQ: Water loss via stomata
5 SQ: Translocation in phloem

Presentations

PN 1 Uptake and transport of water and minerals
2 Loss of water from leaves
3 Stomatal mechanism
4 Cohesion – tension and xylem flow
5 Phloem transport

Video tasks

V 1 Who needs soil? – mineral uptake and nutrition in plants
2 Problems with water – plant water relations
3 Phloem transport

The sequence of the topic is shown in Fig. 4.2. The positions of the informal tutorials vary to some extent with the needs of individual students and by constraints imposed by the nature of the work. Students are not completely free to do what they want as and when they wish. Apart from the need to tackle the work in a logical sequence, some of the practical activities involve using class results so data need to be available to all students at specific points in the sequence.

Students who have been involved in learning in this way have commented favourably on it. Resource requirements are of course demanding, particularly in the extent and quality of the documentation involved. It is essential that such student materials look good, are clear, easy to understand and easy to work from. These aspects are more important in many ways than that the materials are innovative. Hence there is plenty of material available in published courses and resources[44] that can be incorporated into schemes such as the one

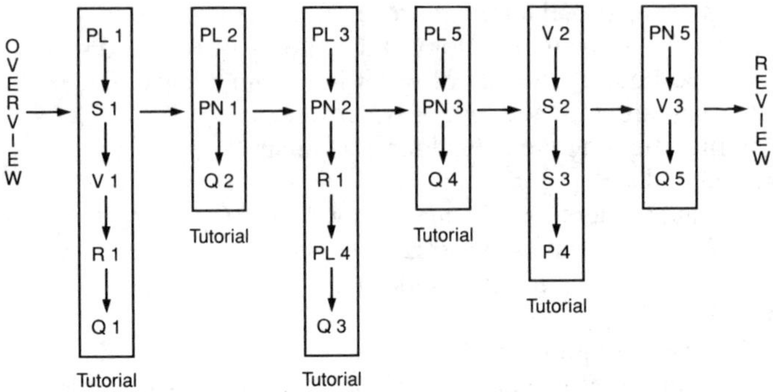

Figure 4.2 Sequence of teaching the topic 'Absorption and transport in plants'

described. Producing such a scheme is, however, a team effort, not something that one biology teacher working on his or her own should attempt. Of course, it makes for a very pleasant start to one's teaching career to begin in an institution in which the bulk of the development has been successfully completed!

And one final thought. Good teaching and learning is founded on identical principles *whatever* ages the learners are!

Notes and references

1. Lunzer, E.A. and Gardner, K. (1979) *The Effective Use of Reading*. London: Heinemann Educational.
2. Sutton, C. (1992) *Words, Science and Learning*. Buckingham: Open University Press.
3. Klopfer, L.E. (1990) Learning scientific enquiry in the student laboratory, in Hegarty-Hazel, E. (ed.) *The Student Laboratory and the Science Curriculum*, London: Routledge.
4. Reading demands can be assessed quantitatively using various 'readability' formulae. For an introduction see Dawson, C. (1981) Appraising a textbook, in Sutton, C. (ed.) *Communicating in the Classroom*. London: Hodder and Stoughton, and *Language in Science, Study Series No. 16*, The Association for Science Education (1980).
5. Still worth looking at, despite the age: Biological Sciences Curriculum Study (1963) *Biology Teachers' Handbook*, p. 39. New York: John Wiley and Sons.
6. For full details see BSCS (1963) *Biology Teachers' Handbook*, op. cit. pp. 201–21.
7. Revised Nuffield Biology (1975) *Text 2 Living Things in Action*. London: Longman for the Nuffield Foundation.

8. Pupils need help along the following lines: If they were given a beaker containing 100 cm^3 of distilled water with 5g of sodium chloride dissolved in it, i.e. a 5 per cent solution, by what two different ways could they make the concentration up to a 10 per cent salt solution?
9. Sutton, C. (1989) Writing and reading in science: the hidden messages, in Millar, R. (ed.) *Doing Science: Images of Science in Science Education.* Lewes: The Falmer Press. Peter Medawar's paper 'Is the scientific paper a fraud?' is reprinted from the *Listener*, 12 September 1963, in Jenkins, E.W. and Whitfield, R.C. (eds) (1974) *Readings in Science Education.* London: McGraw-Hill; and in Medawar, P. (1991) (ed. D. Pyke) *The Threat and the Glory*, Oxford: Oxford University Press.
10. Watts, S. (1980) Children writing, in *Language and Science* (ASE) op. cit.
11. Golby, G. (1976) Science as a process: a change in emphasis, in Schools Council (1976) *Writing in Science, Writing across the curriculum, 11–16 Project*, p. 10. London: Ward Lock Educational.
12. Over-reliance on this has been adversely criticized by HMI, for example in 1979 in *Aspects of Secondary Education in England.* London: HMSO.
13. Bulman, L. (1985) *Teaching Language and Study Skills in Secondary Science*, p. 57. London: Heinemann Educational Books.
14. Davies, F. and Greene, T. (1984) *Reading for Learning in the Sciences.* Edinburgh: Oliver and Boyd.
15. Watkins, O. (1981) Writing: How it is set, in Sutton, C. (ed.) op. cit.
16. See for example, Carré, C. (1981) *Language Teaching and Learning 4 Science.* Slough: Ward Lock Educational; Watts, S. (1980) op. cit.; Lock, R. (1989) Some less usual approaches to pupil work in school biology lessons. *Journal of Biological Education*, vol. 23, no. 2, pp. 142–6.
17. Sutton, C. (1992) op. cit.
18. See Carré, C. (1981) op. cit. for a fuller analysis, and Darke, J. (1980) Children talking, in *Language in Science* (ASE) op. cit.
19. One piece of research has shown that talk between pupils in groups occurs for a longer period in the absence of the teacher and it is richer in ideas and speculation. See Hornsey, M. and Horsfield, J. (1982) Pupils' discussion in science: a strategem to enhance quantity and quality. *School Science Review*, vol. 63, no. 225, pp. 763–7.
20. Adelman, C. *et al.* (undated) *Implementing the Principles of Inquiry/Discovery Teaching Some Hypotheses*, Ford Teaching Project Unit 3, Cambridge Institute of Education.
21. For further discussion on the dynamics of small groups see Barnes, D. and Todd, F. (1977) *Communicating and Learning in Small Groups.* London: Routledge and Kegan Paul.
22. See Dallas, D. (1980) *Teaching Biology Today*, pp. 72–7. London: Hutchinson.
23. Solomon, J. (1993) *Teaching Science, Technology and Society.* Buckingham: Open University Press.
24. The most useful single set of resources for such areas is the SATIS material published by the Association for Science Education. (SATIS,

Science and Technology in Society).

25. See Solomon, J. (1993) op. cit.; Solomon, J. and Swift, J. (1990) Talking about kidney transplants. *Journal of Biological Education*, vol. 24, no. 1, pp. 27–31; and in *School Science Review*, vol. 72, no. 261, pp. 29–34.
26. Examples are to be found in the following resources: *Science in Process*; *Suffolk Coordinated Science*; *Salters' Science*; the SATIS materials; *Better Science No. 4*, *Better Science: A Directory of Resources* (SSCR).
27. A recent report of the results of a school-wide survey of pupils' attitudes to different teaching styles (Christofi, C. and Davies, M., 1991, Science through drama, *Education in Science* No. 141) showed that 70 per cent of pupils liked drama as a way of learning.
28. See for example the materials produced by the Sheffield Biotechnology Group (1987) published by ASE and the descriptions of their use in Henderson, J. and Knutton, S. (1990) *Biotechnology in Schools – a Handbook for Teachers*. Buckingham: Open University Press.
29. *Better Science*, No. 4 p. 11, op. cit.
30. See for an example Somers, J. (1992) Theatre in education in science. *School Science Review*, vol. 73, no. 265, pp. 129–31.
31. Solomon, J. (1963) op. cit., p. 45; and Solomon, J. (1991) *Exploring the Nature of Science*. Glasgow: Blackie and Sons.
32. Henderson, J. and Knutton, S. (1990) op. cit.
33. Ellington, H., Addinall, E. and Percival, F. (1981) *Games and Simulations in Science Education*. London: Kogan Page.
34. Dussart, G.B.J. (1984) Snakes and Ladders as an all-purpose teaching game. *School Science Review*, vol. 66, no. 234, pp. 96–8.
35. Barlex, D. and Carré, C. (1985) *Visual Communication in Science*. Cambridge: Cambridge University Press.
36. For a short review of these developments, see Billings, D. (1993) Science at ages sixteen to nineteen, in Hull, R. (ed.) *ASE Science Teachers' Handbook (Secondary)*, pp. 292–305. Hemel Hempstead: Simon & Schuster.
37. Two of the best known are the Wessex Project (details obtainable from the Wessex Office, c/o Holyrood School, Chard, Somerset, TA20 1JL) and the Cambridge Modular A level Science Project (details from The Secretary UCLES, 1 Hills Road, Cambridge, CB1 2EU).
38. Waterhouse, P. (1990) *Flexible Learning An Outline*, p. 72. Bath: Network Educational Press.
39. Probably the best known and most widely used example of a curriculum incorporating these elements in A-level biology is the Advanced Biology Alternative Learning (ABAL) project developed by an ILEA project team directed by Denis Marshall. *ABAL* (1983) is published by Cambridge University Press as ten units, not all of which are still available.
40. For further details see Waterhouse, P. (1990) op. cit.; and TVEI (undated) *Developments 10: Flexible Learning*, The Training Agency, Moorfoot, Sheffield, S1 4PQ.
41. Waterhouse, P. (1991) *Tutoring*. Stafford: Network Educational Press.

42. Waterhouse, P. (1990) *Classroom Management*. Stafford: Network Educational Press.
43. Mountford, D. (1992) *Flexible Learning Post Sixteen Biology*. Wessex INSET.
44. Materials to support flexible learning in some areas of biology have been produced by the Wessex Project and are held in their Materials Bank. A catalogue can be obtained from the Wessex Office (see ref. 37).

CHAPTER 5

Use of living organisms in biology teaching

It is almost trite to say that since biology is the study of living things, we should expect to see living organisms employed in its teaching in schools. However a recent survey of the use of living organisms in secondary schools has concluded that,

> There are now significant pressures operating against the widespread use of living material in school.[1]

These have resulted in a *decline* in the use of such material despite exhortation to the contrary from organizations such as the Association for Science Education (ASE) and the Institute of Biology (IOB). The most significant pressures reported by teachers in this survey were:

1. that the National Curriculum gave too little emphasis to the use of living material,
2. pressure of time,
3. financial considerations,
4. shortage of technicians,
5. adverse pupil reactions to the use of animals,
6. the proliferation of safety requirements and recommendations.

It is undoubtedly true that pupils are much more ready than they were formerly to voice their thoughts and feelings about the use of living organisms, not only in schools, but also in research and farming.[2] Activities and publications of various animal rights organizations have polarized opinions about dissection into whether it is 'right' or 'wrong' and in some ways this has deflected consideration away from the educational reasons for conducting such procedures. Far less dissection goes on in schools now and it is possible for students to follow a full A-level course and a university degree in biology without doing any whole animal dissection.

Considerable confusion and misinformation exists, even among experienced teachers, about what should be done, what might be done and what cannot be done in schools involving the use of living organisms and material of living origin (i.e. bits of animals such as lungs, hearts, eyes, etc.). In this chapter I will attempt to clarify some of these issues, to indicate accessible sources of further information, provide details of some useful resources and to place in context some of the issues raised as constraints by teachers.

Legal aspects

Currently, the essential starting point for considering legal aspects of the use of animals and plants in schools is the document, *Animals and Plants in Schools: Legal Aspects* (Administrative Memorandum No 3/90, Department of Education and Science, 31 August 1990). The purpose of this memorandum, which has been circulated to all schools, is to clarify the legal position for schools arising from a number of Acts, but particularly from the Wildlife and Countryside Act, 1981 and the Animals (Scientific Procedures) Act, 1986. The key points that could have implications for teachers and which are detailed in the memorandum are summarized below.

1. Dissection of whole animals is *not* illegal provided procedures for humane killing are employed.[3]
2. The Abandonment of Animals Act, 1960 has implications for the care of animals during weekends and holidays.
3. Schools should not allow certain wild animals to be brought in. Some of those listed are relatively unlikely, e.g. crocodiles and alligators, but badgers (alive or dead) come within this category (Badgers' Act, 1973).
4. All wild birds, their eggs and nests (while in use) are protected (except for thirteen 'pest' species). Eighty-two species of birds are listed as being specially protected and a licence is needed from the Nature Conservancy Council to approach or photograph these birds, if near their nests.
5. Other animals protected by the Act include red squirrels, otters and *all* bats, as well as some reptiles, amphibians, fish, insects, spiders and snails (all relevant species are named). This does *not* prevent bringing small quantities of frog or toad spawn into school and studying its development. Good conservation reasons demand that such material be returned as soon as possible, preferably to the place where it was collected.
6. It is an offence to damage, destroy or obstruct access to the place of shelter or disturb in the shelter *all* bats, great crested newt, otter, red squirrel and natterjack toad.
7. Imported species should not be released or allowed to escape into the wild.
8. There are a number of aspects of the legislation that relate to field work.
   (a) It is illegal to capture shrews without a licence. Hence the use of Longworth mammal traps is illegal unless they are

modified by drilling a 13-mm diameter escape hole for shrews in the side.

(b) It is illegal to pick, uproot or destroy any specially protected plants (listed in the memorandum) or to uproot *any* plant grown in the wild.

9. The following procedures are *illegal* in schools (under the Animals (Scientific Procedures) Act, 1986).
   (a) decerebration of an animal by pithing;
   (b) the carrying out of the gastrocnemius nerve/muscle preparation on amphibians, unless carried out after humane killing;
   (c) the injection of hormones to cause spawning in *Xenopus* (African clawed toad), unless this is carried out for a husbandry purpose;
   (d) nutritional experiments involving restricted or excessive diets, although the unusual investigation of plotting the growth rate of a mammal on a normal diet is perfectly acceptable;
   (e) an experiment involving the tossing of a young mouse from hand to hand;
   (f) anaesthetizing an animal (except for humane killing by administration of a lethal overdose);
   (g) any regulated procedure on a tadpole once it is capable of independent feeding; although a tadpole could be expected to have reached this stage two days after hatching, teachers should note that this could be earlier or later, depending on variations in temperature; and
   (h) the examination of living chick embryos after the halfway point of incubation.

Further comments on many of these issues are to be found in ASE (1993), CLEAPSS (1989) and Lock (1989).[4]

Other safety and quasi-legal issues

There are a number of other procedures involving living material or material of living origin on which the DES (now the DfE) has given advice to schools, usually after consultation with the Health and Safety Executive or the Department of Health. In some instances this advice does not prevent the use of that material in schools (although new teachers should err on the side of caution) but individual local education authorities (LEAs) have, in some cases, imposed additional restrictions that must be followed. Copies of such LEA regulations should be filed in schools or they will be available from LEA science advisers.

Blood

Taking samples of blood from teachers or pupils should not be done in schools because of the risk of contracting hepatitis or AIDS. Animal blood, obtained from an abattoir, may be used as a substitute. An anticoagulant should be used to prevent clotting, e.g. a 10 per cent solution of sodium citrate in the ratio of 1 vol. citrate : 10 vol. blood. The citrate solution should be placed in the collecting jar before the blood is added. A sample of blood treated in this way can be kept in a refrigerator for up to two weeks. Care needs to be taken in the use of this blood, for example avoiding direct skin contact with it and mopping up spills with a 10 per cent hypochlorite solution.

Saliva

Contrary to popular opinion, the DfE has not recommended that the use of saliva should cease. However, in most schools of my acquaintance, solutions of commercial bacterial or fungal amylase are used instead as 'artificial' saliva. If saliva is to be used, pupils should collect their own and be responsible for cleaning the equipment they use.

Human cheek cells

The DfE has advised against the practice of taking human cheek cells, advice that has been challenged by the IOB, which regards it as being perfectly safe to take such samples using cottonwool buds. Again, local authority advice should be sought. The demise of the use of cheek cells has probably aroused more concern and annoyance among teachers than most other restrictions. However, the results of using cheek cells were always distinctly unimpressive, and if the temptation to use cheek cells as examples of a 'typical' animal cell has now been removed the advice may prove to have been a blessing in disguise! Alternatives which have been suggested include: scraping a sheep's windpipe; cells from inside a fish's mouth; sticking Sellotape to a well-washed inner wrist before removing it to a slide; and a squashed follicle from a recently plucked hair.

Bovine eyes

The most recent addition to the list of advice from the DfE (February, 1993) relates to the dissection of cows' and bullocks' eyes. Although there is no evidence that BSE (bovine spongiform encephalopathy) can be transmitted to humans, since expert opinion identifies a remote theoretical risk from such dissection, it is best avoided. The circular concerned suggests using anatomical models. It makes no reference to the use of sheeps' eyes. Since scrapie, the disease in sheep that is similar to BSE in cows, has been known for a very long time and has never been implicated in transference to humans, it is assumed that the use of sheeps' eyes is acceptable.

The COSHH regulations

You are likely to have seen something of the operation of the COSHH regulations as a student. As a teacher they take on a slightly different perspective. Some of the requirements are an administrative and legal minefield. Fortunately most of the organizational procedures will have been done for you and a good deal of useful advice is available.[5] Your school science department will have established clear procedures by which it can be shown that the department acts upon the general risk assessments adopted by the LEA. The ASE's Laboratory Safeguards Sub-Committee strongly advises school science departments to adopt the procedure of annotating all departmental documents, e.g. syllabus, schemes of work, work sheets, teachers' and technicians' notes, etc., in order to point out hazards, safety precautions and other requirements. Nevertheless, it is your individual responsibility to understand the implications of risk assessments and for implementing them with your classes.

The regulations relate to the use of chemicals classified as harmful, toxic, very toxic, irritant and corrosive. Obviously a very wide range of chemicals and biochemicals that are used in topics in school biology fall into these categories. Microorganisms also fall within the category of harmful substances and risk assessments are required on them. However they present slightly different sets of problems and they are discussed separately in a later section of this chapter.

Flammables and oxidizing agents are not covered by COSHH regulations but since they pose some of the most frequent and obvious safety problems, the ASE has recommended that teachers treat all chemicals as if they were covered by the regulations. The teacher is required to analyse the procedure they are proposing to use with their classes against the following list of questions:[6]

1. Is this activity educationally necessary?
2. Is there an alternative, less hazardous, substance or procedure?
3. Is it a teacher demonstration or pupil practical work?
4. What is the age and experience of the pupils?
5. How will pupils be warned about the hazards?
6. What is the scale of working?
7. What personal protection or control measures are necessary (eye protection, gloves, fume cupboards, safety screens)?

8. How will residues be disposed of at the end without risk to the technicians?

Which living organisms to use?

There are many published sources of ideas and resources that can be trawled to give you ideas about what materials can be used for what purposes.[7] The selection of such material depends on you and your teaching objectives: what do you want your pupils to know and understand, hence what do you want them to see and to do? In this part of the chapter I will indicate some of the reasons for selecting particular resources for use in schools.

Microbes

Since the Nuffield projects of the early 1960s, in school science in England and Wales, we have begun work on microbiology in years 7 or 8. The justification for this has been the possibility of developing ideas about health and disease, decay, food production and preservation, etc. from interesting everyday examples. Additionally work on microbes has been used to explore a number of important biological principles such as cycles in nature, processes such as growth, respiration and fermentation and some aspects of genetics.

The National Curriculum includes a number of references to these topics in both key stages 3 and 4. Before looking at the sort of work that is done, a word on safety is appropriate. Before any work on microbes is undertaken in school for the first time it is essential that you read the HMI guide.[8] This publication gives details of the vital safety procedures to be employed, summarizes practical techniques to be used and lists appropriate microorganisms for school use as well as those whose use is now considered to be inappropriate. New teachers, even those with limited backgrounds in microbiology, should not be deterred from attempting work with microbes, which can be richly rewarding.

The main safety points for work at this level are as follows:

1. Do not use media that will allow the growth of pathogens. For most purposes use nutrient agar or broth for bacteria and malt extract agar for fungi.
2. Do not incubate cultures at 37°C. Incubate at 25°C for fungi and 30°C for bacteria.
3. Use safe organisms from pure cultures or known safe sources.
4. Do not deliberately place microbes on the skin surface.
5. Do not open plates, which should be sealed with Sellotape after inoculation, unless they have been formaldehyde treated or you

have taken expert advice. Asking the laboratory technician does not come into this category.

6. All cultured material should be disposed of by autoclaving, incinerating or by placing in strong disinfectant, according to instructions.

In an introductory unit of work on microbiology pupils should know and understand that:

1. microbes come in different 'sorts' and are all around us;
2. different microbes require different conditions for growth;
3. microbes can be made to work in the service of humans in a variety of ways;
4. some microbes cause diseases, including human ones.

Some examples of activities that fulfil these requirements follow.

Looking at microbes

Pupils' work should begin by first-hand examination of selected microbes. That they are very small will become obvious as will the need to use a microscope to study them. It may be that this will be the first time that pupils will have used a microscope and hence some brief instruction may be required in the use of the instrument. It is entirely appropriate that they learn how to use the microscope in a real situation rather than by drawing it, labelling the drawing and then putting the instrument away so that they can forget how to use it! One or two ideas follow.

Pupils can undertake microscopic examination of a fresh suspension of baker's yeast and of yeast growing in a culture solution[9] in order to see evidence of budding (stressing that growth and reproduction are characteristics of living things). Similar examination of yoghurt serves as a useful introduction to bacteria. This might be supported by looking at slides previously prepared by the technique of negative staining.[10] Pupils should also be introduced to other fungi. *Aspergillus oryzae*[11] and *Penicillium roqueforti* are useful since they can also be given as examples of microbes that are industrially important. Agar plates can be placed directly on the stage of the microscope and the fungi examined using low power or the Petri dishes can be projected using a film strip projector. A fresh culture of *Neurospora crassa* growing on malt extract agar provides a spectacular demonstration of fast growth by observing the tip of a hypha lined up against the scale of an eyepiece graticule.

Use of microbes in the food industry

A very wide range of different possibilities exists for work in this area. The programme of study for KS3 makes specific reference to the uses of enzymes and microbes in the baking, brewing and dairy industries. Most schools do some work on yoghurt production, wine-making and the manufacture of bread. The well-known procedures, readily adapted to cater for pupils of different abilities, of studying gas evolution from yeast/sugar solution mixtures can form the basis for excellent investigative work. Similarly the extent to which different dough mixtures rise can be studied by placing samples in measuring cylinders, or on a smaller scale, in 1 cm^3 syringe barrels (plunger withdrawn, nozzle plugged with a shortened wooden cocktail stick).

The fungus *Aspergillus oryzae* can be grown on starch agar or on protein agar[12] and demonstration of the digestion of these substrates can lead to further work on the use of fungal and bacterial enzymes in the food industry, for example in altering the sugar content and elasticity of doughs.

Microbes and disease

This has been the aspect of microbiology that has always received the most attention in the school science curriculum. Its place in the National Curriculum, is still secure.

The PoS refers to the way in which the healthy functioning of the body may be disrupted by, among other things, bacteria and viruses, and how the body's defences may be enhanced by immunization and the use of medicines. In the general introduction to the PoS we also find:

> Pupils should be given opportunities to develop their knowledge and understanding of the ways in which scientific ideas change through time. They should study the life and work of some famous scientists and the development of some important ideas in science.

The work of Pasteur, Jenner, Lister, Koch, etc. can be drawn on to good effect in this context. Quality control procedures in the food industry and food preservation technology can also be incorporated into this work.

Testing for bacterial contamination of different types of milk or of samples of pasteurized milk kept for different lengths of time using the dyes resazurin or methylene blue provides pupils with a first-hand example of a quality control procedure used in their local creamery. The action of disinfectants and antibiotics can be demonstrated by applying filter paper squares or discs impregnated with these substances to 'lawns' of the harmless bacterium *Bacillus subtilis* growing on nutrient agar. Pupils can investigate a range of commercially available disinfectants, diluted according to the manufacturer's instructions, toothpastes and mouthwashes that are claimed to be bacteriocidal.

Robert Koch's so-called 'postulates', which represent the model procedure by which microbiologists still investigate the cause of 'new' diseases, involve the following steps:

1. the identification, by microscopic examination, of a potential pathogen within a diseased organism;
2. isolation of the pathogen through continuous culture;
3. infection of a healthy host by inoculation from culture;
4. re-isolation of the same organism in culture after the onset of the disease in the inoculated host.

This sequence can be followed by pupils acting as real disease detectives using *Monilia fructigena* the fungus that causes brown rot disease in fruit. The activity can be made more or less open by varying the approach. The stages for an open investigation are:

1. Examination of rotten and good apples to identify a likely pathogen. Microscopic investigation of the diseased tissue shows the presence of 'threads' (fungal hyphae).
2. Infected tissue is inoculated on to malt extract agar plates. Incubation at 25°C for several days results in the development of an obvious fungus. This can be subcultured if desired.
3. The fungus is introduced into a healthy host as follows. A sound apple is surface sterilized. A wound is made with a sterile scalpel and some of the potential pathogen culture is prodded into the cut. The apple is placed in a polythene bag in a warm place for several days. An appropriate control procedure is also carried out. The two apples are compared and one is seen to be showing evidence of the disease.
4. The fungus is cultured again on agar by taking samples from the newly infected apple.

Apart from the extensive practical activities that can be undertaken, a variety of different teaching approaches, including drama and role play, can be employed.[13] Some excellent video material is also available.[14]

Plants

There are a number of important issues associated with the use of plants in teaching biology.

1. Some aspects of the biology of plants are of central importance in the 11–16 curriculum.
2. Many pupils feel that work involving plants is far less interesting than that with animals.
3. Pupils encounter learning difficulties with certain plant topics, e.g. osmosis, photosynthesis, food webs, etc.

The first point means that we cannot duck the use of plants in our teaching despite the second point! Because of the second point we have to use material that is as stimulating as possible. The third point is taken up in the next chapter.

CLEAPSS[15] has produced an excellent compilation of the ways in which plants can be used in biology teaching and of further references on a variety of topics involving plants. It is pointless to produce a similar list here. A further but briefer account is given by Lock.[16]

In this he lists the following five plants, which he regards as being indispensable in school: *Elodea* sp. (Canadian pond weed); *Pelargonium zonale* (geranium); *Impatiens* sp. (busy lizzy); *Tradescantia* sp. (wandering jew); *Lemna* sp. (duck weed). Whilst not disagreeing with that selection, we all have particular procedures to which we are attached. Here are some of mine.

1. Banana. The browning reaction seen when the flesh of fruit such as apples is exposed to air is used as the basis for a variety of investigations (such as factors that cause it, accelerate it or prevent it). Banana fruit or extract can also be used; the reaction can be rapid.[17]
2. *Buddleia* sp. Because the leaves of this plant are comparatively frost-hardy it is useful for transpiration experiments at 'difficult' times of the year. In early summer, willow or rosebay willowherb also provide excellent material for use in potometers.
3. Buttercup (*Ranunculus* spp.) This is a fascinating plant that can be used for a wide range of studies, particularly involving variation (e.g. in numbers of floral parts).[18]
4. *Crocus* sp. An extract can be obtained from the purple petals of spring flowering varieties that acts as an excellent natural indicator.[19]
5. Garden pea (*Pisum sativum*). Useful material for teaching genetics and supplied in kit form;[20] the tall/dwarf and round/wrinkled varieties are well known. Some of the leaf mutations are interesting and well worth investigating for teaching, particularly the variety *argenteum*. The epidermis is easily removed from leaves and is good material to use for work on stomata.[21]
6. Garlic (*Allium sativum*). Cloves suspended on cocktail sticks with their bases just touching water produce roots in twenty-four hours or so that are ideal for studies of mitosis. Wild garlic (*Allium ursinum*) inflorescences about 1 cm long, collected in spring as soon as they emerge above ground level, provide excellent material for studies of meiosis.[22]
7. Mung beans (Chinese bean sprouts). If you have forgotten to put off seeds and the class is tomorrow (germination, respiration, growth) don't worry, these will germinate and give you radicles 1 cm long if put in a warm place overnight!
8. Onion (red-skinned). The cells of the *outer* epidermis of the scale leaves of these bulbs have red cell sap. They are excellent for

introductory work on cell structure and for demonstrating plasmolysis and recovery.[23] Such bulbs are obtainable from greengrocers and supermarkets throughout the year. Onion sets may also be used. Substitute material that is nearly as good: epidermal peels from rhubarb petiole (seasonal) or from petioles, stems or leaves of the houseplant *Setcreasea purpurea*.

9. Radish. Very useful for germination and growth studies with young children and as an introduction to variation. Useful cultivars: Scarlet Globe (crimson, round); French Breakfast (red and white, intermediate in shape); Icicle (white, very long); Yellow Gold (yellow, intermediate).

Fast plants

Recently the SAPS project[24] has introduced rapid-cycling brassicas, *Brassica rapa*, into school biology in Britain. These plants, which complete their life history (seed to seed) in 35 days, represent some of the most exciting possibilities for biology teaching seen in recent years. And pupils will love them! Currently they are being used for studies in growth and development, photosynthesis, mineral nutrition, plant response and genetics, but new ideas are emerging continuously. The plants require, for their growth, continuous illumination from a light bank of fluorescent tubes. Full details of these exciting developments can be obtained from Richard Price, Director of the SAPS project. Attendance at one of the SAPS in-service courses is a very convenient way of becoming familiar with these materials.

Animals

The compendious sources listed earlier that provide useful information about plants in biology teaching can also be trawled for ideas relating to the use of animals. As before, it is not the intention in this section to duplicate existing sources of such information, but rather to highlight some of the issues that should be considered when contemplating these resources and selecting from them.

Clearly animals can be used in a wide variety of different ways in schools, including functioning as pets. Restricting their use to that of 'moving wallpaper', however, has little to commend it. Small mammals most frequently come into this category, although they can of course be used for a variety of legitimate educational purposes (see below). With the constraints of limited laboratory assistance (especially over weekends and holiday periods), as well as shortage of space and finance, any living animal material must be easily maintained (almost self-maintaining!) and be worth keeping in that its contribution to the quality of teaching is great. In Tables 5.1 and 5.2 I have listed some of the animals that have been most useful in my teaching over the years (still useful in the National Curriculum); hence they are usually present in my teaching laboratory and I find myself recommending them repeatedly to PGCE students. I have divided them into semi-permanent inhabitants of the animal room or laboratory and those which are bought in or collected as required and maintained for short periods of time such as a few days or weeks.

Table 5.1 **Animals held in a permanent collection**

Animal	Suggested teaching use
Mammals	
Mouse (*Mus musculus*)	Behaviour experiments including learning, reproduction, genetics
Mongolian gerbil (*Meriones unguiculatus*)	
Amphibians	
African clawed toad (*Xenopus laevis*)	Capillary circulation in webs between toes
Fish	
Guppy (*Lebistes maculatus*)	Behaviour, adaptation, genetics
Arthropods	
Brine shrimp (*Artemia salina*)	Kept as eggs and hatched when required: life history, behaviour, locomotion[25]
Stick insect (*Carausius morosus*)	Life cycle (including parthenogenesis, moulting, growth), behaviour, locomotion, feeding energetics
Meal worm (*Tenebrio molitor*)	Behaviour (particularly food preferences), life history, gaseous exchange experiments
Flour beetles (*Tribolium castaneum, T. confusium*)	Behaviour, life cycle, genetics
Molluscs	
Garden snail (*Helix aspera*)	Food preferences, life history, sensitivity and behaviour, locomotion (ideal animals for very young pupils – easily handled, slow-moving)
Giant land snail (*Achatina fulica*)	
Annelids	
White worms (Enchytraeidae)	Microscopic examination to show peristalsis in the gut[26]

Table 5.2 **Animals maintained for short periods of time**

Animal	Suggested teaching use
Arthropods	
Freshwater animals	
Asellus spp. (water louse)	Blood system ('open' system, heart action)
Gammarus spp. (fresh water shrimp)	Behaviour (particularly substrate/flow rate preferences)
Mayfly and stonefly nymphs	Ventilation, orientation experiments, adaptations
Woodlice	
Oniscus sp.	
Porcellio sp.	Behaviour experiments (particularly humidity gradients, etc. in choice chambers)
Locusts	
Locusta migratoria	
Schistocerca gregaria	Life history, feeding, flight studies, testis squashes, ventilation, spiracle movements, etc.
Cabbage white butterfly (*Pieris brassicae*)	
Fruitfly (*Drosophila melanogaster*)	Genetics, mating behaviour
Blow-fly larvae (maggots) (*Calliphora* sp.)	Behaviour (particularly phototaxis)
Annelids	
Earthworms	Locomotion, food preferences, behaviour
Tubifex sp.	Blood system (red pigment dissolved in plasma)

Material from the abattoir or butcher

The dissection debate does not usually arise when using material from the butcher or abattoir since the animals involved have already been killed as a source of human food. Nevertheless any issues that are raised should be explored sympathetically, and religious observances and sensibilities should be scrupulously respected.[27]

Some examples of the teaching uses of this sort of material follow.

1. Lungs: the lungs of a sheep, with the windpipe attached (usually with the oesophagus still present, but not the heart, liver and diaphragm, which cost more) can be examined, inflated with a foot pump and the airways explored by dissection. Ask for a 'pluck'.
2. Sheep's head: skinned then bisected vertically can be used to show the brain, jaw muscles and dentition.
3. Kidneys: lambs' kidneys can be cut in different planes to reveal the details of gross structure (see also Chapter 6).
4. Pig's feet: (not 'trotters', which include the next joint up and are more expensive to buy). These are skinned and dissected to show ligaments, tendons and details of the structure of a joint to great effect.
5. Chicken/turkey feet: pulling along the tendons to make different digits move.
6. Bones: fresh bones from a cow; particularly useful are femurs and hip joints. Long bones can be sawn in different planes – include a section of a femur head.
7. Hearts: pig, sheep or cow to explore details of structure including valves, thickness of atrial and ventricular walls, etc.
8. Eyes: structure by dissection. Also deep-frozen eyes can be cut in different planes (whilst frozen) using a hack-saw, then allowed to thaw. See earlier details on bovine eyes (p. 67).

It is particularly important that material like this is presented appropriately to pupils and that they are given an opportunity to opt out of watching the demonstration or of participating in the group activity if they so desire. Usually when invited to sit at the back to do some alternative work, you will find that their curiosity will overcome earlier scruples and that they become enthusiastic participants in the activity. The key to successful presentation of such materials is the matter-of-fact approach taken by the teacher. The material should be cleaned up beforehand, for example by washing it in a dilute solution of disinfectant. It should be covered by a cloth or laboratory paper or otherwise kept out of sight until the demonstration is about to start and the pupils told what they are going to see and do. Pupils will become involved quickly and the level of interest shown will be very high. Teachers using such materials for the first time (the second, really, since it will have been tried out beforehand) will need to be prepared to field a stream of questions such as 'What animal is it from? Is it bigger/smaller/different from a human's? Where did you get it? What are you going to do with it afterwards?'

As an example, here is how a sheep's pluck might be used as part of a lesson with year 9 or 10 pupils. The material is presented as described above. The teacher points out the windpipe, asks for comments on the different appearances of windpipe and oesophagus, on the colour, texture and shape of lungs (already addressing one popular misconception: that lungs are like two inflated rubber balloons). Next, the lungs are inflated using a foot pump fitted with rubber tubing that is fed well down the windpipe. The pupils are amazed by the different appearance and feel of the inflated lungs (NB the meat inspector will almost inevitably have slashed one lung, but that doesn't matter). The teacher allows the pupils to examine the larynx, to be surprised that the vocal cords are not like taut guitar strings ('How do they work then, Miss?') before cutting along the length of the windpipe, taking in the C-shaped cartilages on the way ('Just like the hose of a Hoover'). Pupils see, and feel, the slimey lining of the windpipe and they begin to speculate about the significance of the tunnels they can see leading off the windpipe. Following these deeper into the tissue of the lungs, they find that the branches become finer and more numerous but where *do* they lead? A whole series of questions will have been raised by now and need to be followed up, but before that some consolidation is needed by having the pupils note their own observations, in their own words of course.

Dissection of whole animals

Even at A level no pupil should be forced to dissect a small mammal or to watch such a dissection if they would prefer not to do so. Below A level, and with the proviso that pupils can opt out of the activity, teachers may decide to conduct demonstration dissections to teach: the structure of the gut, the positions and relationships of heart and lungs, the layout of the organs of the uro-genital system, etc. Teachers must decide whether such demonstrations are the best way to address their teaching objectives and meet the needs of the pupils. Alternatives to whole mammal dissections exist in models, photographs, computer simulations and videos.[28] Purchase of a video camera is within the financial reach of most science departments or schools nowadays. As is often the case, an in-house video made in the department is probably going to be a more valuable teaching resource that commercial material because you know what it is you want your pupils to see. I use video materials of this sort that show:

1. progressive unravelling of a rat's gut alongside a metre rule;
2. inflating a rat's lungs *in situ* in the thorax with air via a catheter inserted into the trachea and fitted to a 20-cm^3 plastic syringe;
3. how a silicone rubber internal cast of rat lungs is made;
4. a preparation involving attaching a rat's windpipe to a small manometer and the subsequent pressure changes that occur in the intact thorax when the diaphragm is pulled downwards and allowed to recoil;

5. the phases of wing movements of an otherwise immobilized locust when viewed under a stroboscope.

Apart from the obvious advantage of not having to sacrifice an animal each time, these videos involve procedures that are time-consuming to set up and which, in some cases, are difficult to demonstrate to a whole class.

Field work and the teaching of ecology

I have included a short section on this topic here because ecology occupies a unique position in the teaching and learning of biology. It has an integrative nature, drawing separate strands of biology such as physiology, behaviour and genetics together and, not least, because there are difficulties involved in teaching it.[29] The Cinderella-like place of ecology in the school science curriculum was analysed in some depth in 1979 by P. R. Booth,[30] then a member of HMI. Many of the issues raised then have been addressed many times subsequently,[31] but they have not gone away. Booth's comment, that 'it seems doubtful that much ecology is taught or learnt in schools' was supported by the identification of a number of issues that contributed to that situation. Among these were:

1. the elusiveness of some of the principles involved (leading to lack of clarity about what should be learnt and understood);
2. difficulties and confusion about how to teach ecology including:
   (a) treating it as a separate topic rather than as an integral part of the course,
   (b) lack of definition of clear objectives, leading particularly to use of techniques for their own sake,
   (c) attempting to take on over-ambitious tasks with too many variables involved;
3. that teachers' confidence was low in tackling this work, with identification posing particular problems in this respect.

The effect of these various constraints, compounded by the nature of examination syllabuses and the opportunity for candidates to avoid questions in the examination that tested ecology, resulted in the situation quoted earlier by Booth. Where ecology was taught it tended to involve sixth form students undertaking a week's study in a field centre of some sort, where the teaching was done by staff of the centre rather than by the pupils' own teachers. Harper in a provocatively titled paper, 'Why not abolish ecology?'[32] suggested that ecology should not be taught as a topic, unconnected to other areas of the syllabus, but that it should be integrated as a theme running through the course and dealt with when appropriate, e.g. work on food chains would follow photosynthesis. This suggestion had echos of the earlier approach taken in the Nuffield O and A level biology courses,[33] where not only was a thematic approach taken, but also considerable use was made of artificial habitats such as damp bread, lumps of meat, cleared patches of soil and also of microcosms such as small tanks of pond water. The rationale for this approach was that it was easier to study, say, colonization and competition at first hand over a few days in the laboratory rather than, perhaps literally, putting pupils in a situation where they might not be able to see the wood for the trees! Or in other words, to employ smaller scale investigations with clearly defined objectives and a limited range of potential variables!

Coming up to date, the present National Curriculum places considerable emphasis on field work at all four key stages, as is shown in the following extracts from the programmes of study:

Key stage 1	'Pupils should study plants and animals in a variety of local habitats, for example, playing field, garden and pond.'
Key stage 2	'Pupils should explore and investigate at least two different habitats and the animals and plants that live there.'
Key stage 3	'Pupils should study a variety of habitats at first hand . . .'
Key stage 4	'Pupils should make a more detailed and quantitative study of a habitat, including the investigation of the abundance and distribution of common species, and ways in which they are adapted to their environment . . .'

Further details are provided in the National Curriculum about the particular concepts that are to be developed through this work, supported by some examples. The extracts quoted above do, however, indicate some of the problems that teachers are having to face in developing work in these areas. What habitats are available for first-hand exploration that are not only proof against the invasion of year groups of perhaps 200 children (not necessarily all at the same time, but seasonally!) but which are also sufficiently rich (both in terms of species to study and problems to investigate) to provide progression when revisiting through successive key stages? Just how many times can you use the playing field or the school pond! Clearly schools will vary considerably in the extent to which they are able to draw on local resources. And then there are the vital issues of health and safety to be considered.[34]

One of the most fundamental issues to be solved is how do you start this work? Almost certainly what *not* to do is to let them loose on an unsuspecting habitat in order to 'see what you can find'. There does seem to be considerable benefit in introducing such work by bringing the habitat into the laboratory. The idea of using a large polythene bag of leaf litter collected from a suitable deciduous wood (an activity that can be done from late May through to autumn when the litter is rich in animals) was mentioned in Chapter 3. There it was discussed as an opportunity to allow pupils to raise questions that could lead to designing and carrying out investigations, essentially physiological or behavioural (feeding, moving away from light, humidity preferences, etc.), building towards the concept of adaptation detailed in the PoS for Key Stage 4.

Pupils hand-sort the litter, making collections of the animals as they do so, including using pooters to collect fast-moving ones. The animals are identified using appropriate identification aids.[35] The work can be extended in a number of ways such as by using slightly more sophisticated extraction techniques (Tullgren and Baermann procedures) and growing up microbes from the litter (and hence moving into ideas about decomposition, recycling, pyramids of number, etc.). After these introductory activities it would be profitable to extend the work to examining a real wood if one is nearby. A parallel approach can be taken using a couple of bucketfuls of water taken from a pond or ditch.[36]

The school grounds offer considerable scope for field work and many teachers have done successful work using the school field, surrounding hedges, walls, pavements, tree trunks, and small areas of derelict land and so on.[37]

Profitable work arises from situations where pupils are led to see that certain organisms are unevenly distributed in a particular location. The question 'Why?' follows spontaneously. This raises the important concepts of microhabitat requirements, competition, adaptation and succession. These concepts arise easily in situations where the numbers of different organisms are fairly restricted. Here is an example for which the requirements are minimal – a lawn, an area of less frequently mown grass, e.g. the surrounds of a playing field, home-made half metre square quadrats and a means of identifying lawn plants from their vegetative features.[38] Pupils, working in small groups, are required to sample five quadrats (0.5 × 0.5 m), in each of which they record if a plant is present (*not* the numbers of plants of each species in each quadrat, which clearly causes difficulties, similarly 'grass' is usually excluded from the analysis). Class results taken in two different locations always show dramatic differences in the distributions (Table 5.3).

A number of follow-up investigations can then be undertaken, some of which might involve the identification, control and manipulation of variables, thus addressing some of the higher levels of Sc1. Work of this sort is important in that it is complementary to the essentially correlational nature of other aspects of field work. For example comparing the differences in ground cover at different light intensities in different parts of a wood will lead to the formation of a hypothesis that pupils are unlikely to be able to test given the usual constraint of time and opportunity.

Table 5.3 **Distribution of plants in twenty-five quadrats**

Plant found	*Frequency of occurrence in 25 quadrats*	
	Lawn	Un-cut field
Broad-leaved plantain	6	3
Dock	2	2
Dandelion	11	8
Creeping speedwell	19	2
Yarrow	1	1
Moss	16	6
Creeping buttercup	2	7
Mouse ear chickweed	3	2
Daisy	23	3
White clover	12	10
Spear thistle	1	0
Sow thistle	1	0
Lesser yellow trefoil	12	0
Pearl wort	1	0
Self heal	5	0

An important extension of this principle of the uneven distribution of organisms is to base work on comparing habitats that show sharp differences in populations of animals and plants. Comparisons of polluted and unpolluted freshwater streams or of fast and slow moving streams are examples that lend themselves to this approach. After initial (and fairly standard, but that does not make them unimportant) sampling techniques,[39] questions are raised that demand further experimental work in the laboratory and field. This is a good yardstick by which to evaluate field work: it goes beyond collecting, identifying, making lists, proceeding to raising questions, generating and testing hypotheses both in the laboratory and in the field. These possibilities are particularly important in meeting the requirements of A-level syllabuses. Many of the A-level boards require an individual project or assignment and field-related topics clearly offer enormous scope in this context.

Notes and references

1. Reiss, M. J. and Beaney, N. J. (1992) The use of living organisms in secondary school science. *Journal of Biological Education*, vol. 26, no. 1, 65.
2. Millett, K. and Lock, R. (1992) GCSE students' attitudes towards animal use: some implications for biology/science teachers. *Journal of Biological Education*, vol. 26, no. 3, 204–8.
3. For details of humane killing of small mammals see Wray, J. D. (1974) *Small Mammals*, Schools Council Educational Use of Living Organisms Project. London: English Universities Press.
4. See Lock, R. (1993) Use of living organisms, in Hull, R (ed.), *ASE Secondary Science Teachers' Handbook*, pp. 179–205. Simon and Schuster; CLEAPSS (1989) *Laboratory Handbook*, CLEAPSS School Science Service, Brunel University, Uxbridge; Lock, R. (1989) Investigations with animals and the Animals (Scientific Procedures) Act 1986. *School Science Review*, vol. 71, no. 255, 74–5.
5. See CLEAPSS (1989) op. cit. and Borrows, P. (1993) in Hull, R. (ed.) op. cit.
6. Borrows, P. (1993) op. cit.
7. See CLEAPSS (1989) op. cit. which has an invaluable section for biology teaching; see also Kelly, P. J. and Wray, J. D. (eds) (1975) *The Educational Use of Living Organisms: A Source Book*, particularly pp. 123–33. The other volumes in this series, published by English Universities Press for the Schools Council are equally valuable.
8. DES (1990) *Microbiology; an HMI Guide for Schools and Further Education*, second impression. London: HMSO. See also the relevant section in CLEAPSS (1989) op. cit.
9. Dissolve 2 g of malt extract in 100 cm^3 of warm distilled water. Place in a 250 cm^3 flask fitted with a cottonwool (non-absorbent) bung. Autoclave.
10. See any standard laboratory text on microbiology methods.
11. Supplied as a kit, with an instruction leaflet by Philip Harris Ltd.
12. Brown, C. R. (1989) Protease activity of *Aspergillus oryzae*. *School Science Review*, vol. 71, 81.
13. As a starting point see Henderson, J. and Knutton, S. (1990) *Biotechnology in Schools, A Handbook for Teachers*. Milton Keynes: Open University Press.
14. For younger pupils use the Scientific Eye programme 'Microbeasts and disease'.
15. CLEAPSS (1989) op. cit. pp. 540–1570.
16. In Hull, R. (1993) op. cit.
17. Investigations on the enzyme responsible for the browning reaction, appropriate for A level, are described in Brown, C. R. (1983) Banana polyphenol oxidase. *School Science Review*, vol. 64, 690–5.
18. The best account of this is: Cunnell, G. J. (1964) The morphology and anatomy of the buttercup flower, in *The Science Master's Book, Series 4/part 3 Biology*. London: John Murray. It may still be on some preparation room shelves.

19. Extracted in ethanol, the indicator keeps indefinitely.
20. Supplied by Philip Harris Ltd.
21. See Ambrose, M. J., Brown, C. R. and Taylor, P. N. (1993) New pea genes for the classroom. *School Science Review*, vol. 75, no. 271, 87–90.
22. They can be preserved in 70 per cent ethanol or deep frozen. For staining procedures see Nuffield A-level Biological Sciences Laboratory Guide.
23. Details are in ref. 21.
24. Details from Richard Price, Director, SAPS, Homerton College, Hills Road, Cambridge, CB2 2PH.
25. Ward-Booth, K. and Reiss, M. (1988) *Artemia salina*: an easily cultured invertebrate ideally suited for ecological studies. *Journal of Biological Education*, vol. 22, no. 4, 247–51.
26. Starter cultures can be obtained from most aquarists. Semi-permanent cultures are set up in damp peat (or substitute) in a sandwich box. They should be fed occasionally by adding small pieces of bread soaked in milk to the medium. Since they are semi-transparent their gut can be seen clearly under the microscope.
27. Very useful teaching material that presents views about the need for animals in research can be obtained from the Animals in Medicines Research Information Centre, 12 Whitehall, London, SW1A 2DY. A video, 'Life on the Line', is also available free of charge.
28. For a list of such resources see RSPCA (1984) *Alternatives to dissection* Horsham: RSPCA. The whole area of dissection is explored in Dixon, A. (ed.) (1988) *Issues Surrounding the Use of Animals in Science Lessons: a Resource Pack*. Hatfield: ASE.
29. See Dowdeswell, W. H. (1981) *Teaching and Learning Biology*, ch. 7. London: Heinemann.
30. Booth, P. R. (1979) The teaching of ecology in schools. *Journal of Biological Education*, vol. 13, no. 4, 261–6.
31. Recently by Openshaw, P. H. and Whittle, S. J. (1993) Ecological field teaching; how can it be made more effective? *Journal of Biological Education*, vol. 27, no. 1, 58–66.
32. Harper, G. H. (1982) Why not abolish ecology? *Journal of Biological Education*, vol. 16, no. 2, 123–37.
33. See particularly Monger, G. (ed.) (1975) *Revised Nuffield Biology: Text 3: Living things and their environment*. Longman.
34. See CLEAPSS (1989) op. cit. p. 1702; DES (1989) *Safety in Outdoor Education*. London: HMSO; Institute of Biology (1990) *Safety in Biological Field Work: Guidance Notes for Codes of Practice*, 3rd ed. There will also be guidance from your LEA about taking pupils out of school.
35. By using pictures, and progressing to keys, see Chapter 7 for difficulties pupils have in using keys.
36. Taking the appropriate safety procedures of covering recent cuts, washing hands after the activity and warning not to put fingers near faces – all to

guard against the possibility of contracting Weil's disease. For further details see CLEAPSS (1989) p. 1706.

37. For many sources of ideas for this work see Collins, M. (1984) *Urban Ecology*, Cambridge: Cambridge University Press.
38. See Bishop, O. N. (1983) *Adventures with Small Plants*. London: John Murray.
39. For details and further references see Bennett, D. P. and Humphries, D. A. (1983) *Introduction to Field Biology*, 2nd edn. London: Edward Arnold; Williams, G. (1987) *Techniques and Fieldwork in Ecology*. London: Bell and Hyman.

CHAPTER 6

Teaching about the structure of living organisms

Structure and function

That structure and function are complementary is not only one of the fundamental tenets of biology but also one of the oldest, going back to Aristotle. There are two important and interrelated ideas here: what an organ (or other structure) *does* depends on its structure and, secondly, that its function can be identified from a consideration of the structure. William Harvey wrote in 1628:

> The motion and circulation of the blood are confirmed from the particulars apparent in the structure of the heart, and from those things which dissection unfolds.[1]

Here we have not only a philosophical position but also a methodological one relating to the role of dissection! More recently, and at an entirely different level of biological organization, one of the most famous pieces of scientific writing of the twentieth century, states:

> It has not escaped our notice that the specific pairing we have postulated immediately suggests a possible copying mechanism for the genetic material.[2]

Different context, same point: given this structure, this is what it might do. These are instances of how great biologists have worked in the past, but have they anything to do with teaching pupils? I am certainly not suggesting that pupils' learning should be based on an inductive approach by which the teaching begins with a tabula rasa and proceeds to build by making and collecting first-hand observations, from which subsequent theories are fashioned.[3] There are many different ways of illustrating the structure of living organisms, including models, film, video and computer simulations. Nevertheless there are three main reasons to turn to details of real structure from time to time:

1. Looking at real objects aids understanding.
2. Such understanding is part of personal knowledge and an important aspect of a general education.
3. Many biological structures have great delicacy and beauty and hence a strong aesthetic appeal to pupils.

Try looking really carefully (e.g. using a low power stereo-microscope) at the 'seed-end' of a dandelion fruit and its parachute, or at the awn of a fruit of the wild oat *Avena fatua*. Then try to answer the questions that arise about the functions of 'hairs' on fruits or the 'rifling' on the awn or the subsidiary projections on the parachute. Better still, ask a class of pupils!

Two examples of teaching structure–function relationships

Lungs

In the last chapter, part of a lesson was described in which the teacher used a sheep's pluck and from which the question arose 'Where do all those little tubes lead?' Establishing the nature of the fine structure of lungs will be part of everyone's lesson on 'gaseous exchange', 'diffusion in the lungs' or whatever you decide to call it. Very soon we will be talking about large surface area, thin walls, extensive capillary beds, etc. For pupils to *understand* these very difficult ideas it is essential that they have a good idea of the three-dimensional nature of the air sacs and their relationship to the airways. A conventional microscope preparation such as a stained transverse section of lung takes you some way towards this, but not far enough. We have made extensive use over the years, to some effect I think, of two different preparations: internal casts of small mammal lungs[4] (rat in practice) and thick sections of dried lungs[5] (rat again). These resources have allowed exploration of a number of ideas but, particularly, in building up three-dimensional perceptions of the structure of the lungs. I will return to this point later in the chapter.

Kidneys

The following sequence shows how an investigative approach can be taken in a topic in which some aspects of function can be inferred by considering appropriate aspects of structure. A year 10 class, for which this approach has been planned, will have access to prior knowledge of the details of the blood circulation and their everyday knowledge that urine is produced by humans and is discharged periodically from the bladder via the urethra.

Resource 1 is a dissected rat with the gut removed to display the kidneys, ureters, bladder, urethra, renal artery and vein, aorta and posterior vena cava. Pupils are invited to examine this overall layout: blood in, blood out, two fine tubes leading to a bag, and consider what might be happening to produce a liquid that is, obviously, not blood.

Resource 2 is an intravenous ureogram (begged from a hospital). Pupils are informed about the technique used to produce the X-ray: a radio-opaque material is injected into a vein, following which the kidneys are photographed. Video material showing this is also available.

Resource 3 the filtering idea that pupils will advance can be underlined by a simple but spectacular demonstration. A marker injected into the pelvis of a kidney *in situ* in a freshly killed (or deep-frozen and thawed) rat will pass down the ureter into the bladder.[6]

Resource 4 addresses the question of the nature of the filtering mechanism that we have inferred is operating. A corrosion cast preparation is examined in which the renal artery of a sheep's kidney has been injected with latex or silicone 'rubber'. Preparations of this sort reveal that blood vessels appear to terminate in knot-like structures (glomeruli).

Finally pupils need some input via photographs and prepared slides about the structure of nephrons. Ideas about the filtration mechanism are developed from a working model[7] and the function of nephrons from data such as that referred to in Chapter 4.

I hope that you can see how such resources would meet the claim made in the introduction to this example. Perhaps one word of caution is also necessary; this approach could not be adopted for every organ or system. For example, other people looked at the structure of the heart for at least 1000 years before Harvey did, without drawing the conclusions for which Harvey is justly remembered. And it will be the same for your pupils. Presented with a sheep's heart and invited to investigate its structure and then decide how it works will have just one outcome – disaster! Before pupils examine a heart they must know something about it: that it has four chambers, how they are interconnected, in what directions blood flows in the heart and around the body, including the lungs. This calls into play other resources such as photographs, drawings and diagrams.

Pictures, diagrams and other two-dimensional representations

School textbooks make extensive use of a wide variety of images – sometimes well over half the pages of GCSE or A-level books carry photographs, line drawings or diagrams, often in colour. These devices gain attention, motivate and focus attention on particular details and through them pupils learn about the structures presented. As with most things concerning teaching, it is not quite as simple as that. A fairly typical example from a school textbook is shown in Fig. 6.1. The structural details seem to be fairly straightforward, to an experienced

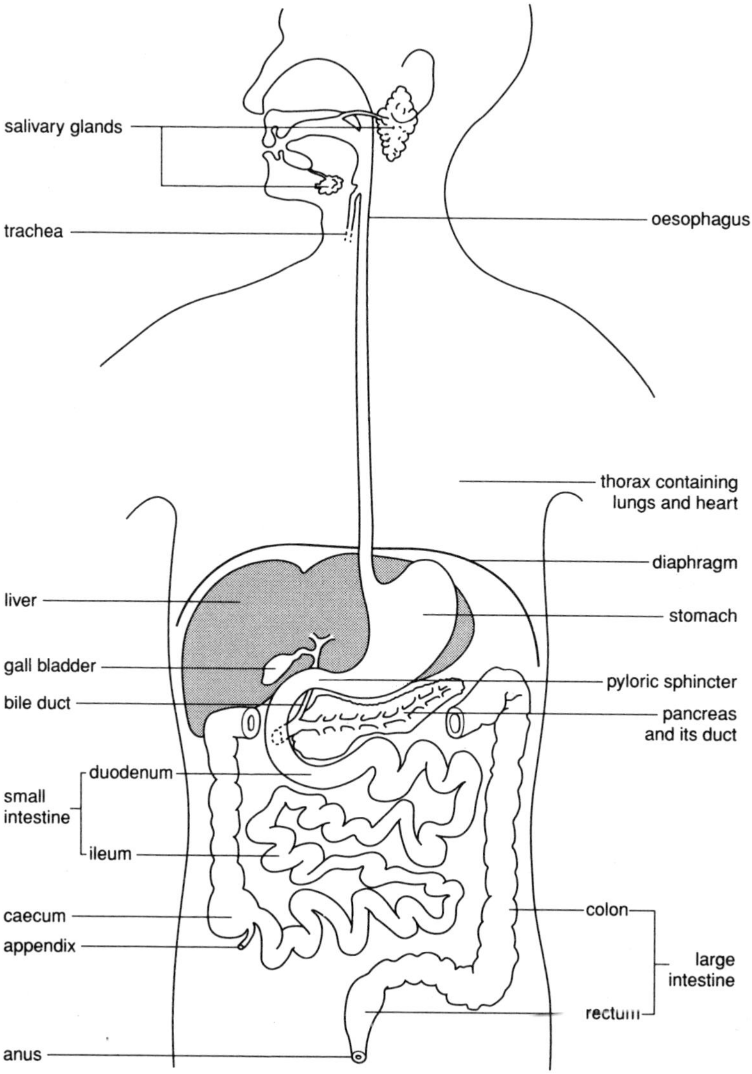

Figure 6.1 A diagram of the human digestive system. The horizontal part of the large intestine has been cut away and the small intestine considerably shortened.

observer. But what is it like seen through the eyes of a thirteen-year-old meeting it for the first time? Some of the issues/confusions that might arise are:

Which parts are surface views (of whole organs) and which are sections? (*to the reader*: does it say?)
Which organs are hollow, which are solid?
What is the meaning of the different forms of stippling?
What do the dotted lines mean (trachea, the unlabelled entry of bile/pancreatic duct into the duodenum)?
What do the two concentric circles refer to (on structures labelled 'anus', 'colon' and 'caecum')?
Is the bile duct in surface view or a section? And the pancreatic duct?
Are the gall bladder, stomach and oesophagus *on* or *in* the liver? (Hence, is this organ shown in section?)

So how accurate would their conceptions of these structures be if they were left to their own devices with this as the sole teaching resource, 'Make your own notes from fig X page X'. I am exaggerating to make the point; however, try similar analyses on some examples drawn from textbooks that your pupils use, among which I suggest you include the heart and the ear!

Perhaps, when reading the last section, your attention was drawn to features of the diagram that you might not have noticed in a cursory examination. Two important related points about the observer emerge from this, as Barlex and Carré put it:

> Perception depends upon what we know already, and we do not see things as they are, we see them as we are.[8]

This will be picked up again later in the chapter.

But, from what was said earlier, learning is also determined by aspects of the structure of the diagram itself. In an important review of the role of pictures in learning biology, Reid[9] points out that ambiguity arises in the learner when a number of shapes are presented together as in the earlier example. Such ambiguity is heightened by some of the conventions that are used extensively in biology texts, for example in showing cut surfaces when attempting to illustrate sectional views. In one study[10] eleven-year-olds failed to identify the cut surface in five out of six examples taken from biology textbooks. This was despite their ability to do so in an illustration of a cut ball. In these sorts of illustrations, recognizing the cut surface is a perceptual preliminary to appreciating the three-dimensional structure of the object. This is one example of a range of visuo-spatial skills that are important in biology and which pupils develop to greater or lesser extent over time.

Three different visuo-spatial skills were examined in a Scottish study. These were the ability

1. to visualize a two-dimensional section taken from a three-dimensional structure;

2. to visualize a three-dimensional structure when given appropriate two-dimensional sections;
3. to recognize a structure when its orientation was changed.[11]

Although the spatial tests used in the study involved geometric shapes and everyday objects such as clothes-pegs and rulers, the researchers identified some important implications for biology teaching. If objects are viewed at a different orientation (e.g. under a microscope) to that shown on a blackboard or in a textbook illustration, 40 per cent of twelve year olds and 25 per cent of fifteen year olds may not recognize it. Pupils find it easier to move from three-dimensional to two-dimensional images (ability 1 above) than the converse (ability 2), possibly because they encounter that process more frequently and earlier in their work in science.

Other studies have examined spatial abilities within biological tasks, such as those shown in Fig. 6.2.[12] From this work it was possible to suggest that understanding of three-dimensional structure seemed to be based on two different skills: abstracting sectional shapes (example A in Fig. 6.2) and appreciating the spatial relationships of three-dimensional structures, seen in different sectional planes (example B in

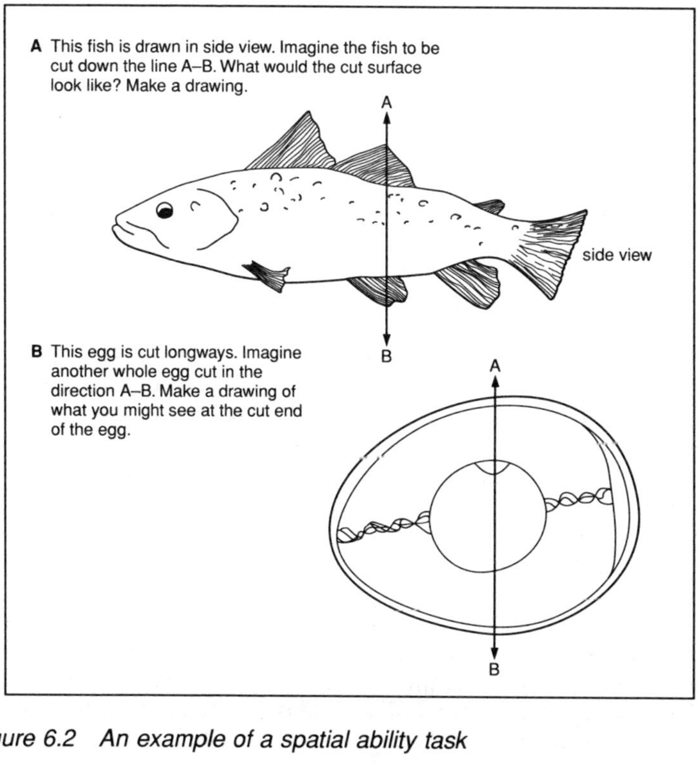

Figure 6.2 An example of a spatial ability task

Fig. 6.2). Some pupils seemed to be competent in both skills; in others performance was more variable including competence in one skill but not in the other. In commenting on the findings of this research, the idea was raised that pupils' understanding might be enhanced if they were presented with actual specimens instead of pictures. In a small-scale study one of my students explored this possibility, i.e. that there might be a 'real-object superiority' effect.[13] A test was used in which six items tested abstraction of sectional shapes by the reconstruction of imagined sections (1–6 in Fig. 6.3) and four items tested the appreciation of the relationships of internal parts in imagined sections (7–10 in Fig. 6.3). Groups of pupils were given the test either as drawings or the objects themselves (the red blood cell was presented as a model). Although the numbers of pupils involved were small and hence the results had to be interpreted with caution, an interesting pattern emerged in that pupils' performances were higher when using *drawings* on items 1–6 but the *reverse* was true for items 7–10.

Some interesting common errors were also noted:

1. adding bits of the rest of the specimen to the abstracted section (to form a kind of stereogram);
2. including imagined extraneous features (e.g. bits of gut, blood, etc.);
3. transposing structures to wrong positions or drawing entirely different views.

Implications for teaching

As usual we can make no assumptions about the quality of the learning that takes place; we must be sensitive to the potential difficulties which pupils may experience in areas of learning requiring understanding of structural interrelationships. Opportunities must be found to explore with pupils their thoughts about what they are seeing and the interpretations they make from those experiences. In many situations, but particularly in microscope work, since we do not know what pupils perceive when they look at objects, we must find ways of sharing images with them. The traditional way of doing this in biology teaching is to ask pupils to draw what they can see. Perhaps we make too little use of asking them to describe and possibly to do both. But, in any case what is essential is that these observations are cross-checked with the teacher and pupils are given opportunities to clarify their thoughts about what they are observing. Let us take an example.

A class of eleven-year-old pupils is looking at some onion epidermal cells for the first time. Supported by a blackboard drawing, they are encouraged to look for the cell wall, cytoplasm, vacuole and nucleus. Hence here we are dealing with two different two-dimensional

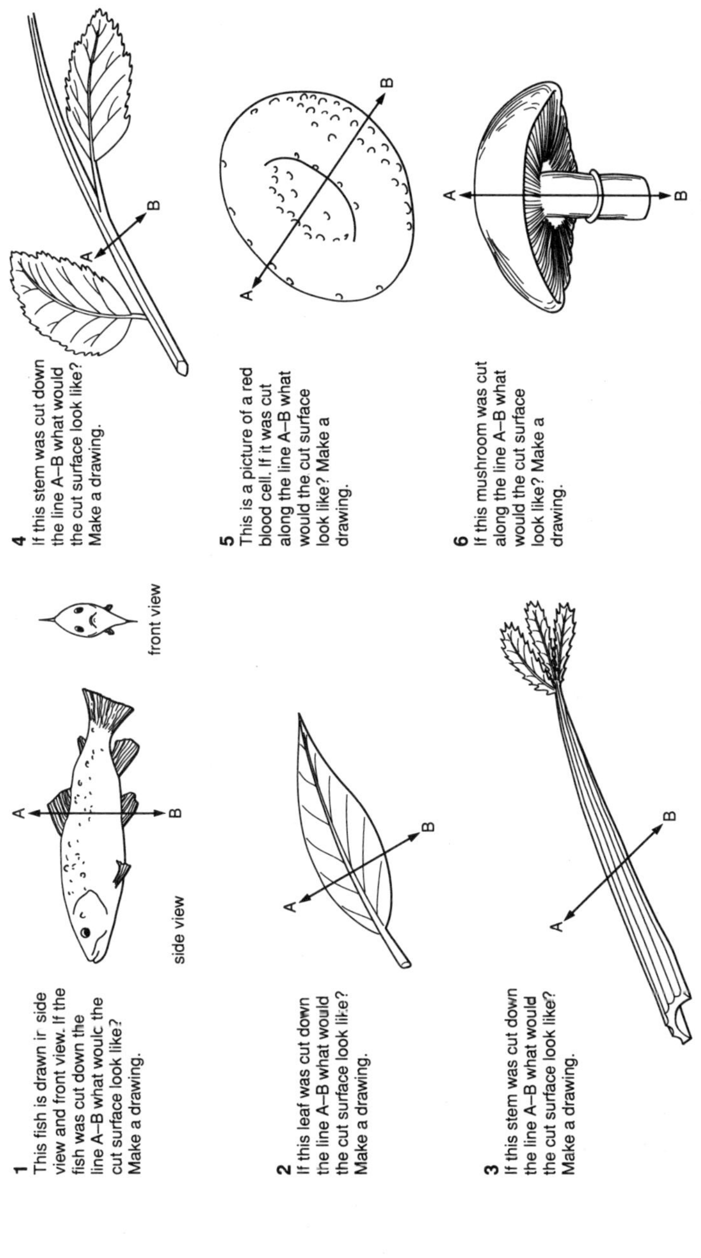

Figure 6.3 A test of visuo-spatial skills in biology (continued overleaf)

7

This egg is cut longways. You can see what it looks like inside. Imagine *another whole* egg cut in the direction A–B. Make a drawing of what you might see at the cut end of that egg.

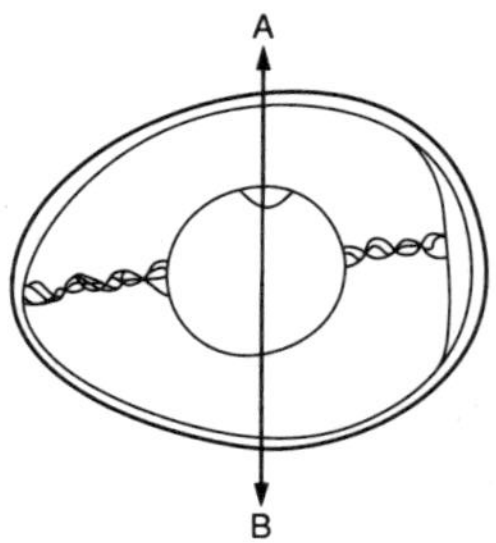

8

This is a picture of an apple cut in half and a view from above the apple. Imagine *another whole* apple cut in the direction A–B. Make a drawing of what you might see at the cut end of that apple.

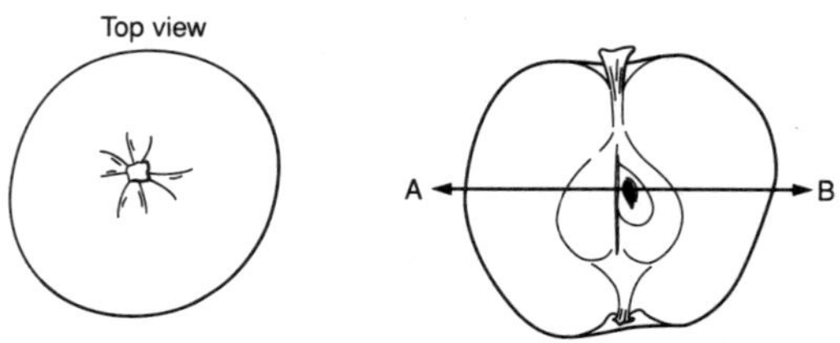

9

This is a picture of a pepper cut in half and a view from above the pepper. You can see what it looks like inside. Imagine *another whole* pepper cut in the direction A–B. Make a drawing of what you might see at the cut end of that pepper.

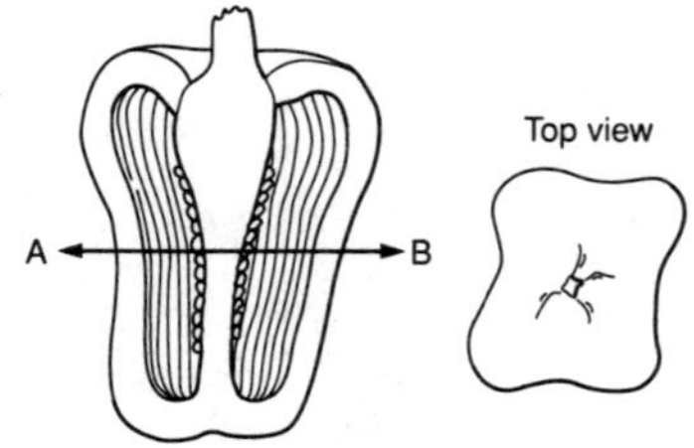

10

This is a picture of a bean cut in half and a view from above the bean. You can see what it looks like inside. Imagine *another whole* bean cut in the direction A–B. Make a drawing of what you might see at the cut end of that bean.

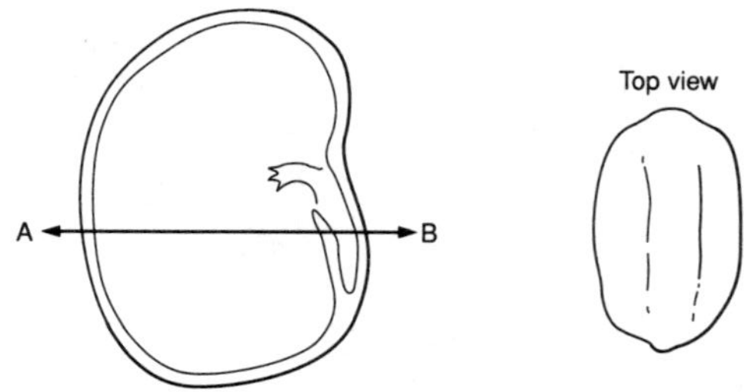

Figure 6.3 (continued)

images, one on the blackboard, one under the microscope, and asking pupils to relate them. Let us introduce a third two-dimensional image: a model cell is placed on the overhead projector and the two-dimensional image on the screen looks similar to the other two (*we* think!). The model cell is a clear plastic lunch box inside which is a polythene bag filled with a solution of a red dye (we are using red-skinned onions for this work of course) which also contains a polystyrene sphere. A teaching approach that takes account of the visuo-spatial points discussed earlier continues along these lines:

Teacher: 'You've been looking at a section of the cell under the microscope, a slice through it because its walls are see-through. What else do we need to know to tell us what the shape of the cell is like?'
Pupil 1: 'What its top is like.'
Pupil 2: 'What it's like end-on.'

The next response of the teacher, how the model cell is used, is obvious. Then:

Teacher: 'Did you see a nucleus in every cell? [Pupils answer No] Why not?'
'What would we have to do to check if there was a nucleus in a certain cell?'

At first the teacher helps the pupils to move from a two-dimensional to three-dimensional perspective, then explores spatial relationships within the cell, moving from three-dimensional to two-dimensional.

In passing we should note how useful the overhead projector is as a teaching device in exploring such areas. Simple models of animal cells can be made from stiff plastic bags filled with water and containing beads of different sizes and one larger sphere. Many other possibilities will occur to you.[14]

Other representational models

A very wide range of models can be used in teaching, both of the commercial variety and home-made ones such as those mentioned in the previous section.[15] The usefulness (and representational accuracy) of many of these will have to be evaluated by you as part of your consideration of what resources you wish to use in your lessons. Keep in mind that most models will be inaccurate in some details and hence a very useful way of summarizing is to ask pupils to identify how the model is inaccurate, e.g. the well-known bell-jar demonstration of lung inflation which I include in my own teaching specifically for this purpose.

Instead of just looking at models, pupils can be involved in constructing them, which is a useful additional way of exploring three-dimensional understanding. These may be fairly simple and the construction task a short one integrated into the lesson, for example asking pupils to make a model red blood cell from plasticine to check on understanding of solid shape. More ambitiously, pupils can be encouraged to take on more demanding tasks such as making models of a leaf, a cell or a 'phage particle'.[16] Time may not be available to construct some of these more elaborate models (and you may not feel that such time expenditure is justified anyway) but we all can use everyday objects as effective models or as analogues, e.g. a fried egg as a squamous epithelial cell, a veal ham and egg pie as a cell, or cutting drinking straws in different planes to explore the nature of the many tubular structures that crop up all the time (xylem, phloem, arteries, veins, nephrons, etc.). Be imaginative!

Pupils observing

There are many reports of the difficulties pupils have in making and recording observations when confronted with the task for the first time. A much-quoted example is that of pupils looking down the microscope at some onion cells and drawing some very scrappy things that bear little of the expected detail, and including air bubbles and the margin of the field of view. They do not filter out from the wide range of possible observations that can be made those which are relevant. Observation is sometimes regarded as a general educational skill (perhaps even cross-curricular and transferable between contexts) and the word is used almost synonomously with 'seeing'. *Scientific* observation is something rather different, a point that has been clarified in the writings of the APU science teams, for example:

> The importance of observation in science has led to a considerable amount of philosophical reflection on what the act of observing involves. In the context of science, observation is not restricted to the sense of sight. It includes all the senses, and it is not a passive activity. Merely 'seeing' something could be adequately tested by employing the methods used by opticians. When engaged in scientific observation the observer uses an intellectual framework in order to know what to see; how to look at the world. This framework is acquired through experience and training in the discipline of science.[17]

In other words, observation in science is 'theory-laden' taking us back to the statement earlier in the chapter that 'perception depends upon what we know already'. The implications for teaching are clear. Pupils must be required to make observations within a framework of known concepts and facts. So even in an A-level task, e.g. expecting students to map the distribution of different tissues making up an organ seen in section such as a stem or root, their ability to do so depends on knowing the differences between tissues or that an epidermis is present or absent.

The APU[18] have reported some important findings about pupils' competence in making scientific observations, which are useful to bear in mind:

There is a disparity between what pupils can do and their ability to give reasons for doing so, e.g. in grouping tasks.
Pupils, when asked to identify similarities and differences, tend to imply rather than make explicit, points of comparison involved.
They are content to note the most obvious variables, particularly visual ones and to neglect others (particularly non-visual).
Younger pupils (up to age 13) find it easier to identify and describe differences rather than similarities.
Pupils find it harder to describe events rather than similarities or differences between static objects (or photographs or drawings).

In the next chapter we will look in more detail at some of the other aspects of pupils' difficulties in learning in biology.

Notes and references

1. William Harvey (1628) *On the Motion of the Heart and Blood in Animals.* Worth reading the original text if you have never done so.
2. Watson, J. D. and Crick, F. H. C. (1953) A structure for deoxyribose nucleic acid. *Nature*, vol. 171, 737–8.
3. For analysis of these issues see Driver, R. (1983) *The Pupil as Scientist?* Milton Keynes: Open University Press. Particularly Chapter 1.
4. Hillier, D. V. (1973) Some simple uses of rubber latex. *School Science Review*, vol. 55, no. 191, 302–5; and Brown, C. R., Hillier, D. V. and Rees, A. (1986), More ideas for teaching lung structure. *School Science Review*, vol. 67, 741–2.
5. Gill, B. F. (1981) Teaching lung structure – the use of thick sections of dried rat lung. *School Science Review*, vol. 62, no. 220, 492.
6. Hillier, D. V. (1973) op. cit.
7. See the filtration model in Nuffield Teacher's Guide 2 (1975), page 252 involving the top of a Kipp's apparatus and a perforated polythene bag.
8. Barlex, D. and Carré, C. (1985) *Visual Communication in Science.* Cambridge: Cambridge University Press.
9. Reid, D. (1990) The role of pictures in learning biology: Part 1, perception and observation. *Journal of Biological Education*, vol. 24, no. 3, 161–72; and Part 2, picture–text processing. *Journal of Biological Education*, vol. 24, no. 4, 251–8.
10. Constable, H., Campbell, B. and Brown, R. (1988) Section drawings from

science text books: An experimental investigation into pupils' understanding. *British Journal of Educational Psychology*, vol. 58, 89–102.

11. Macnab, W. and Johnstone, A. H. (1990) Spatial skills which contribute to competence in the biological sciences. *Journal of Biological Education*, vol. 24, no. 1, 37–41.
12. Russell-Gebbett, J. (1984) Pupils' perceptions of three-dimensional structures in biology lessons. *Journal of Biological Education*, vol. 18, no. 3, 220–5 and Russell-Gebbett, J. (1985) Skills and strategies – pupils' approaches to three-dimensional problems in biology. *Journal of Biological Education* vol. 19, no. 4, 293–8.
13. Hogg, M. E. (1991) *Visuo-spatial Skills in Biology*, unpublished B.Sc. dissertation, University of Hull.
14. For a useful account of the overhead projector see Barlex and Carré (1985) op. cit.
15. For an account of the use of models and an extensive list of commercial supplies and references see CLEAPSS (1992). *Laboratory Handbook Supplement, Models in teaching biology*, pp. 1529–35.
16. Lock, R. (1991) Creative work in biology – a pot pourri of examples Part 2. *School Science Review*, vol. 72, no. 261, 57–64.
17. Driver, R., Gott, R., Johnson, S., Worsley, C. and Wylie, F. (1982) *Science in Schools, Age 15: Report No 1*. London: HMSO.
18. Schofield, B. (ed.) (1989) *Science at Age 13: A review of APU Survey Findings 1980–84*, see Chapter 7. London: HMSO.

CHAPTER 7

Some ideas in biology are more difficult to teach than are others

Ask experienced teachers to think about their teaching of a range of different topics in biology and they will quickly volunteer their thoughts that some of those topics are more difficult to teach than are others. Many teachers report that feeling of desperation when the pupils just do not seem to be 'with' the teacher in grasping the key ideas involved in the topic. In the last fifteen years or so these intuitive ideas of teachers have been confirmed by extensive research that has revealed that many different aspects of school science do pose problems of understanding for pupils and they are responsible for creating significant barriers to pupils' learning. Clearly two issues of great importance for novice teachers are: which topics pose these problems and what can be done about addressing the difficulties? The questions are deceptively simple; unfortunately finding the answers is much more complicated! This is partly because areas of difficulty are different for different reasons, and hence a single, overall solution to the problems cannot be identified. Nevertheless, the research that has been undertaken has been of great benefit in identifying potential trouble-spots (forewarned is forearmed!) and in some instances has led to the development of some tactics for teachers to use to combat them.

Constructivism and the teaching of plant nutrition

A class of fourteen-year-olds is about to begin a new topic, 'photosynthesis'. They have previously covered the topics of 'respiration' and 'digestion' in which attention has been confined largely to the consideration of the human animal. The teacher begins by inviting the pupils to work in groups to discuss the question, 'How do green plants get their food?', and then to summarize their thoughts on a poster and report their ideas to the rest of the class. Each group is then asked to test their ideas by doing an investigation. What is happening here is a demonstration of a new approach to the teaching of plant nutrition.[1] At the heart of this approach lies the idea that the learner constructs his or her own knowledge and what is constructed depends crucially on what the learner brings to the situation as well as the nature of the learning situations provided by the teacher.

Many pupils, when put into the situation described above, respond to the question posed with answers such as 'from the soil' or 'in the water taken up by the roots'. Clearly as a result of planning and performing a test of these ideas, they will be forced to the conclusion that their original ideas must have been incorrect. The important point about this part of the strategy of the teacher has been in recognizing that the pupils do not begin their thinking from a blank sheet; they carry ideas to the situation gleaned from a range of prior experiences many of which originate from outside school as part of everyday events (watering plants, 'feeding' tomatoes, etc.). These ideas can be held tenaciously and they can function as blocks to the learning of the correct scientific ideas. However learning will take place if such everyday ideas are rejected by the pupils themselves.

The next phase of the teaching is to present pupils with the appropriate scientific ideas about photosynthesis in what is, basically, statements of key facts. Pupils go on to perform appropriate practical activities (such as testing leaves for starch after they have been maintained in different conditions) and relate their findings to the previously presented scientific explanation. Finally, they are encouraged to reflect on how their ideas contrast with their (now rejected) first responses to the initial question.

This approach is very different from the way we conventionally approach the teaching of most school topics in biology. We usually employ a 'guided-discovery' approach in which we encourage pupils to work things out for themselves, as far as possible, by providing them with a series of planned experiences. Some of the problems inherent in this approach have been identified.[2] An example of teaching plant nutrition by a guided-discovery approach can be seen in the *Revised Nuffield Biology* (Text 2, Longman, 1975). The development of the topic occurs as follows (somewhat abbreviated here):

1. Pupils are challenged to consider the apparent paradox that despite carbon dioxide production by the respiration of living organisms the concentration of it in the atmosphere remains roughly constant.
2. Second-hand data, relating to the diurnal variation in carbon dioxide content in air taken from between leaves in a field of long grass, is presented to pupils in order for them to generate the testable hypothesis that 'green plants take up carbon dioxide when illuminated'.
3. Pupils examine the net exchange in carbon dioxide between air and green leaves at different light intensities. Building on their knowledge of respiration, they hypothesize that another process involving exchange of carbon dioxide is at work leading to a net uptake of carbon dioxide in strongly illuminated leaves.

4. Further second-hand evidence for their ideas in (3) is provided by considering the uptake of isotopically labelled carbon dioxide by *Chlorella.*
5. The fact of the near-constancy of oxygen in the atmosphere is presented to the pupils and leads to a sequence similar to (2) and (3) above.
6. The concept of the food chain is introduced to focus on ideas about the sources of food.
7. The conditions under which starch can be found to be present in leaves are investigated by pupils (in light, in the presence of carbon dioxide and in variegated leaves).
8. Pupils deduce that starch is synthesized from carbon dioxide in leaves.
9. A practical demonstration is conducted of the synthesis of starch (more accurately, of a starch-like substance) from glucose-1-phosphate using a starch-free sample of juice prepared from a potato tuber.
10. Respiration and photosynthesis are compared and the relationship between the two processes in the growing plant are established.

The approach looks attractive: it is logical; certainly it offers pupils an interesting variety of learning experiences; and it builds on work previously covered and, presumably, concepts mastered such as 'food' and 'respiration'. So, where are the problems? Let us make a slight digression to take in some research findings.

The Assessment of Performance Unit, in their 1984 survey of fifteen-year-old pupils, included the question 'Grow-tree' shown in Fig. 7.1. Pupils' responses to the question fell into four main groups, the nature and frequencies of which are shown in Table 7.1.

Examples of responses in each of the four main categories are:[3]

1. 'The 250 Kg came from CO_2 and water and was converted by photosynthesis to sugars and then into 'wood'.'
2. 'The extra Kg came from the substances it has taken from the soil and the rain ect [sic] and they have been used to make it grow begger [sic].'
3. 'The extra 250 Kg comes from the things it eats and drinks from the ground. It's just like us eating and getting bigger.'
4. 'As the tree grows it expands in all directions, the trunk gets wider and taller, there will be more branches and leaves, and so the weight of the tree is larger.'

A small tree is planted in a meadow. After 20 years it has grown into a big tree weighing 250 kg more than when it was planted.

Where do the extra 250 kg come from? Explain your answer as fully as you can

....

....

....

....

....

Figure 7.1 'Grow-tree'

Table 7.1 **The percentage of pupils giving different types of response to the question 'Grow tree'**

	Per cent pupils		
Type of response	Total (*n* = 759)	Pupils studying biology (*n* = 355)	Pupils not studying biology (*n* = 406)
Tree grows because it makes tissue from constituents it takes in from its environment	8	12	4
Tree grows because of what it takes in from its environment	26	28	25
Tree grows because it takes in food from its environment: from soil, from sunlight	19	20	18
Tautology – tree grows because it gets bigger (more branches, bigger trunk, etc.)	43	38	47
No response, uncodeable, don't know	4	2	6

Clearly, it is only in the first example in which the key ideas of photosynthesis are used. The very small number of responses of this sort caused surprise and alarm among teachers when these figures were published, particularly since pupils who had opted to carry on with their biology did not fare much better than those who had studied the subject for the first three years of secondary education and had then dropped it. Responses like the second example revealed evidence of a deficiency in learning: even though ideas of uptake were correct any reference to photosynthesis was missing. In the third example a misconception concerning the nature and origin of plant food is revealed. Interview studies carried out with fifteen-year-old pupils[4] have shown that pupils' understanding of 'food' is likely to be in terms of the everyday notion, mentioned earlier, of any material that is taken up by the plant. It is at variance with the scientific ideas of food as material that functions as, or gives rise to, respiratory substrate or providing for new growth. Alternatively pupils conceive food in the way in which we talk about human food, 'it's just like us, eating and getting bigger' and it is necessary 'in order to keep us alive.'[5]

Perhaps after this brief look at some of the APU findings in this area you are beginning to see how the problems that pupils experience are not addressed in our example of a guided-discovery teaching scheme. The issue of what counts as food is treated as being non-problematic; that the synthesis of starch being centrally involved in the process is assumed to be an obvious learning outcome of the sequence. However, as the APU findings suggest, pupils may perceive the role of a white powder that they see in a laboratory bottle labelled 'starch' as being of less than central importance in their thinking about food. Fourteen-year-old pupils find it fairly unlikely that a gas, carbon dioxide, and water when taken in by the plant will eventually give rise to the solid, brown stuff, the wood, of which trees are made![6] Typically, teaching approaches to photosynthesis lack stages 1–6 of the sequence described earlier. The topic begins by considering the conditions under which starch can be detected in green leaves and therefore lacks any logical processing towards key ideas as well as taking no account of any ideas that children bring to the situation.[7]

Logical processing towards the acquisition of key ideas in photosynthesis depends upon pupils having mastered prerequisite concepts such as 'gases', 'energy', 'food'. Such concepts then function as steps of a ladder. In many of these areas, research has revealed that pupils have difficulties in mastering the subordinate concepts and hence progress towards the top of a concept hierarchy is unlikely to be attainable.[8]

Osmosis

A good example of how effective concentration on the learning of pre-requisite ideas can lead to improvements in handling difficult ideas has been seen in approaches advocated for teaching osmosis. Since this topic is not one that children are likely to meet in everyday life, confusion between everyday and scientific meanings cannot be advanced in order to explain the well-documented difficulties that pupils experience in this topic.[9] As a result of their researches in Scotland, Johnstone and Mahmoud have shown that an effective route to learning osmosis can be constructed by making use of the idea that when a solute dissolves in a liquid the solvent takes an active part in the process. The effect of this in, say, a sucrose solution is to reduce the number of 'free' water particles. When a sucrose solution is separated from pure water by a partially permeable membrane, the *relative* numbers of water particles on the two sides of the membrane can be compared, are found to be different and hence diffusion (a phenomenon within the pupils' experience) can be invoked to explain a

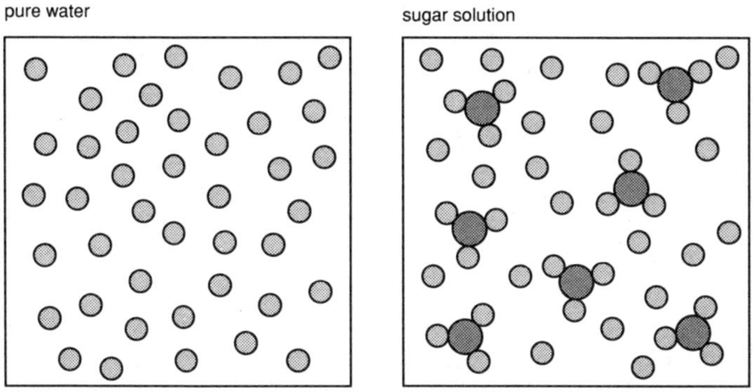

(a) This is 100 per cent free water molecules

(b) Only 50 per cent of these water molecules are free

Figure 7.2 Osmosis

laboratory phenomenon that will previously have been explored by the pupils (see Fig. 7.2).

The important points about this teaching approach are that it *worked* in helping pupils to understand a 'difficult' topic[10], it is based on sound physico-chemical principles consistent with other experiences about the behaviour of particles; and it does not store up problems for later work in that there are any inaccuracies to be corrected.

Children's ideas about 'living'

The biologist's criteria for something being alive, transmitted to pupils via dozens of treasured teachers' mnemonics are, of course: growth, respiration, excretion, reproduction, irritability, movement and nutrition. Recently I was in a science lesson with some eleven year olds. The teacher lit a candle and asked the class if it was alive; about half the class thought it was. She then blew out the candle flame; everyone was now quite certain it was not alive! Similar confusion surrounds children's thoughts about the sun, fire and clouds. Animistic views of this sort have been attributed to children since Piaget's study in 1929[11] in which he showed that a series of stages occurred in the development of the concept of 'living'. Children become better at classifying animate and inanimate objects correctly as they get older but this facility varies greatly. Children can correctly classify animals as being alive more frequently than they can plants. Seeds and embryos are most frequently wrongly classified, for example in an interview study:

'Seeds are dead. When we put them in the soil they . . . begin to live'.[12]

The reasons children give for identifying if an object is alive are diverse, although movement has been shown to be among the most frequently-used criteria.[13] A complicating factor here for the teacher lies in the use of language: there is an everyday meaning to 'living' and 'live' as when we talk about a 'living fire' or 'a live wire'. Clearly if we are to help children to clarify their thinking about these ideas, we must indicate to them if our interest lies, at that moment, in the everyday or in the scientific word. We must construct the intended meaning with them.[14]

Other problems arise in the degree of accessibility of the ideas associated with the seven characteristics of living things.[15] It is within the everyday knowledge and experience of eleven year olds that living things, particularly human ones, move, feed, grow, reproduce and are sensitive to changes in their surroundings. This everyday knowledge can function as a key starting point for an exploration of these ideas. Excretion and respiration, however, present more formidable problems. Not only will pupils be unfamiliar with the words 'excrete' and 'respire', but their knowledge of their own bodies and how it functions does not provide direct access to these concepts for the teacher to use. The distinction to be made between removal of metabolic wastes and that of waste material from the gut is hard-won after many hours of teaching of the relevant areas. Understanding this does not follow from short definitions. Much better is to wait until the appropriate time.[16]

We asked pupils in the first year in secondary school whether they had heard the word 'respiration' before. A tiny minority of those asked had done so, but only in the context of 'artificial respiration'.[17] Fourteen and fifteen year olds who had been taught the topic (in successive years in the case of the older pupils) still encountered problems in disentangling ideas on breathing and respiration. We gave them the following multiple-choice question as part of a test:

Respiration is
A breathing
B filling the lungs with air
C the way energy is released in living things
D what you do when trying to save someone's life

Their responses to it were as follows:

	Per cent of pupils selecting each response	
	14 year olds	15 year olds
Answer	(n = 82)	(n = 60)
A	41	22
B	18	5
C	32	73
D	9	0

The confusion here, we felt, was not attributable to inadequate teaching since the teachers involved were aware that the objectives of the study included analysis of pupils' understanding of the concept 'respiration', and the lessons were audio-taped and checked for their content accuracy and appropriateness. The figures given above do give some cause for cautious optimism. Careful structuring of the work and an awareness of the kind of difficulties pupils might encounter produced an improved level of learning.

Respiration and photosynthesis

You will not be surprised to learn that much confusion occurs in children when they are asked to consider possible interactions between these two processes. Many children think that plants do not respire at all. You see why: if respiration is confused with breathing then clearly green plants do not do it! Another commonly held view is that plants respire only in the dark (or at night), in the light they photosynthesize. These sorts of difficulties may at least partly be attributable to presentation of the two processes as 'opposites', for example in summary equations or in tabulated comparisons.

A well-known school experiment involves investigating colour changes of hydrogen carbonate indicator caused by the gaseous exchange of green plants under different light conditions (Fig. 7.3). The usual teaching approach first involves establishing that the colour change that occurs in the dark tube is due to carbon dioxide production by the leaf. Hence the colour change that occurs in the fully illuminated tube is due to a net uptake of carbon dioxide, i.e. that the rate of uptake exceeds that of carbon dioxide evolution. A far simpler,

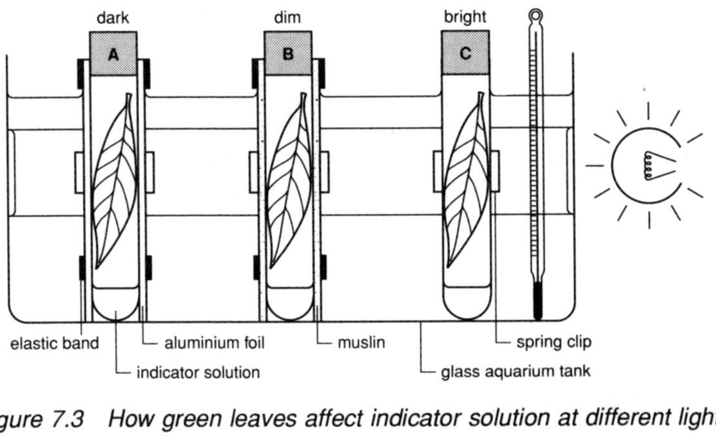

Figure 7.3 How green leaves affect indicator solution at different light intensities

and to pupils a far more satisfactory explanation, is that in the light carbon dioxide production ceases, to be replaced by a process of uptake!

A further difficulty occurs as a result of the description of the green plant as a 'primary producer' and of animals as 'consumers' (primary, secondary, etc.). This terminology occurs in too many school biology textbooks to list. Green plants are, however, producers and consumers. They 'consume' (in respiration, etc.) some of their photosynthetic product. Trained biologists can cope with the shorthand form: plants are producers. The ideas that we should be trying to establish with pupils is that a green plant experiences a net gain in dry mass in a typical twenty-four-hour time period (although not necessarily in those words).

Cognitive development and curriculum demand

So far in this chapter I have given examples of topics in school biology that pupils find difficult and I have provided evidence for the statement I made at the beginning of the chapter, namely that a single method of tackling all of these problems has not been identified. Among reasons given for the learning difficulties, I have mentioned the confusion that can arise between everyday and scientific meanings of words; that ideas brought to science lessons from life outside school may conflict with those which the teacher is attempting to promote; inadequate structuring and preparation of ideas may occur; and inappropriate presentational strategies may be offered by the teacher. Dealing with the curriculum in this way is piecemeal. Michael Shayer has argued persuasively that a model of the learning process must be found that 'can be applied over a wide range of individual learning experiences'.[18] He considers that the key features of such a model are that it addresses with equal power both the learning material and the pupils' thinking processes. With his co-workers he has provided such a model based on Piaget's stage theory, the outlines of which are summarized below.

Pupils in secondary school are likely to show two major types of thinking. Younger and less able pupils are *concrete operational* thinkers. Older and abler pupils can deal with abstractions and complex problems, which are the distinguishing strategies of *formal operational* thinkers. The development of thinking in children follows the sequence of stages shown below:

1	pre-operational,
2A	early concrete operational,
2B	late concrete operational,
2B/3A	transitional,
3A	early formal operational,
3B	late formal operational.

All children pass through the sequence (although not all attain, even when adult, the thinking abilities that characterize 3B) but they do so at different rates and hence at different ages. Shayer has developed science reasoning tasks by means of which the stages of cognitive development of individual pupils can be assessed. Population norms have been developed from this work (Fig. 7.4).

The other strand of this work has involved the development of so-called curriculum analysis taxonomies. The first of these describes, for each stage, the characteristic of children's thinking that is applicable to their general understanding of science. The second taxonomy concentrates on the thinking demands of particular science activities and again assigns these to a stage. Finally, examples of topics selected from the three sciences are analysed.

The two strands of the work come together in that they offer the chance to identify mismatches between the conceptual demand of the lesson, topic or curriculum and the cognitive development of the pupil. Clearly if the demand of the topic (in stage terms) is greater than the development of the pupil, the pupil will not be capable of understanding the work. Let us look at an example.

Originally appearing in the second-year material of the Nuffield Biology course (for year 8 pupils), and extensively adopted by teachers subsequently, was a simulation of one of Pasteur's experiments (Fig. 7.5).

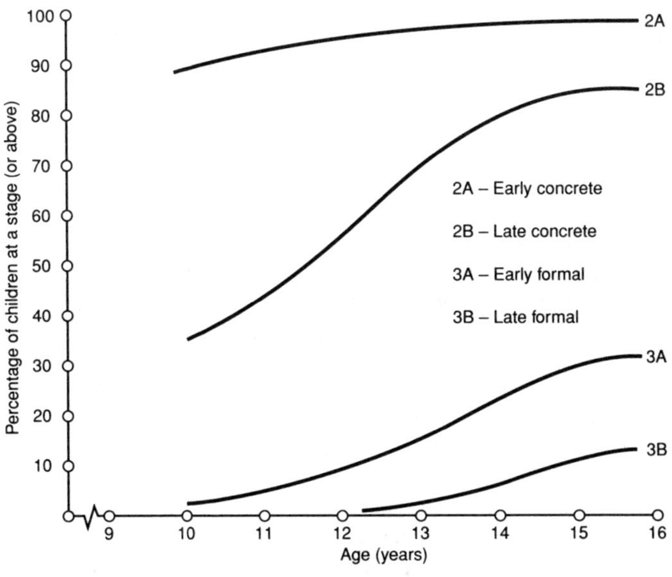

Figure 7.4 Proportion of children at different Piagetian stages in a representative British child population

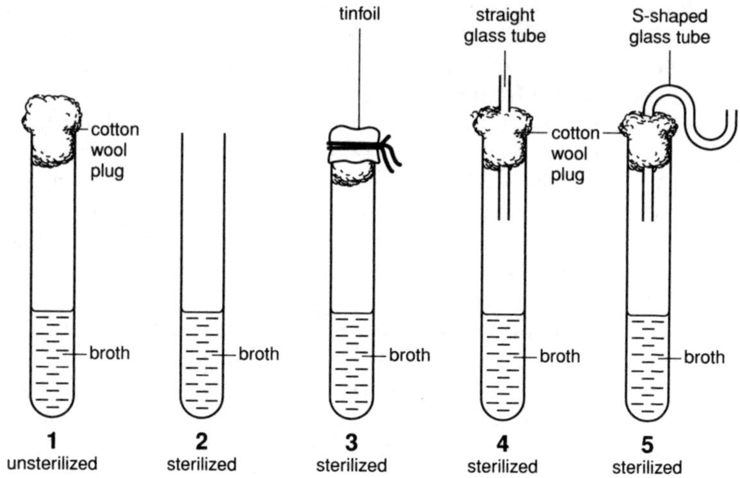

Figure 7.5 Discovering where bacteria come from

The expectation is that pupils will deduce that substantial growth of microbes occurs only when the nutrient broth is already previously contaminated or if it becomes contaminated subsequently by airborne microbes (and hence they will appreciate the significance of the S-shaped tube). Applying part of one of the taxonomies to this procedure we find that the cognitive demands of the experimental set-up are as shown in Table 7.2.

If you refer to the likely distribution of stages of development in a typical mixed ability class (Table 7.3) clearly the demands of this experiment (the number of variables involved) outweigh the cognitive level of a very large number of the class.

As long ago as 1974, Ryman suggested that the difficulties twelve year olds encountered in the classification of living things might be due to the fact that most pupils of that age would be at the concrete level and the tasks that they found difficult were dependent, for their successful completion, on abstract thought.[19] Shayer and Adey's work confirms this suspicion. Interestingly this particular aspect of biology teaching has more recently received the attention of workers in New Zealand employing a constructivist model rather than a Piagetian one.[20] Young pupils equate the term 'animal' with the farm or jungle mammals with which they are most familiar (in books or as toys) from an early age. Other animals are excluded from this grouping, e.g. 'No, that's not an animal, that's an insect.' Typically the ability to classify

Table 7.2 **A curriculum analysis taxonomy**

Type of problem	*2A early concrete*	*2B late concrete*
2.5 Control of variables	Can reject a proposed experimental test where a factor whose effect is intuitively obvious is uncontrolled, at the level of 'that's not fair', but fails to separate variables and so eliminate one. 'Fairness' may also be applied in the sense of giving every factor an equal chance, e.g. 'Slower runner should be given shorter distance'	Will usually vary more than one factor in each experiment, and often varies other factors to test the effect of a given one
	3A early formal	*3B late formal*
	Sees the need to vary one factor at a time and can suggest experimental tests to control for factors explicitly named. May fail to control factors that are not perceptually obvious. Fails to develop a strategy based on a feeling of the system as a whole. May not see the point of having an experiment without a factor present to see if it is a variable	Sets up suitable experiments to economically control factors and eliminate ones that are not effective, and can apply 'all other things equal' strategy to multivariable problems. More sophisticated biological experiments possible including interaction effects. Appreciates the impossibility of controlling natural variation, and so the need for proper sampling

Table 7.3 **Range of ability in an 'average' class**

	Age		
Stage	*11+*	*13+*	*15+*
<2A	2	0.5	0
2A & 2A/B	15	7	3
2B & 2B/3A	11	17	17
3A	2	4	6
3B	0	1.5	4

Source (Tables 7.2 and 7.3: Shayer, M. and Adey, P. (1981) *Towards a Science of Science Teaching.* Heinemann Educational Books.

using class inclusion strategies, (e.g. a bird is so because it has feathers, it is also an animal . . .) is demonstrated at the late concrete operational stage. To the constructivist, however, the problem exists in the confusion that occurs between the everyday and scientific usage of the term 'animal'. Different diagnoses necessitate different treatments! The constructivist's advice to teachers is to help children to contrast the different uses of words and to appreciate that there are these contrasts that pupils will encounter. The proponent of stage theories sees a solution, not so much in looking for better teaching methods, but in improving the match between demand and ability to meet it and, more recently, in identifying ways of intervening in children's development and so to accelerate it.[21]

Genetics, natural selection and evolution

Genetics

Many studies of students' learning have involved asking them to rank topics by their perceived difficulty. Quite often the results have been surprisingly different from the perceptions the students' teachers hold! The topic of genetics, however, is consistently located near the top of these 'ladders' of difficulty by pupils and teachers. A typical quotation, from an A-level student illustrates the point:

> it just appears to be like maths to me. I just don't like maths at all – numbers! I would sooner have something visual.[22]

These topics occupy a central role in the biology curriculum at all levels. The fact that students find them difficult is cause for great concern. What sort of demands do we make of pupils?

The National Curriculum for science (in England and Wales) includes in the work expected at Key stage 4, i.e. the fourteen to sixteen age range, the following:

> Understand the principles of a monohybrid cross involving dominant and recessive alleles. (Sc2/8c)

I suggest that the following concepts must be *mastered* by pupils if this statement of attainment is to be achieved: gene, allele, chromosome, homologous chromosomes, separation of homologous chromosomes in meiosis, segregation of alleles into different gametes, relative frequency of different gamete genotypes, rules of probability in zygote formation, that chance accounts for ratios of phenotypes such as 3 : 1. The essential component of *understanding* the principles of a monohybrid cross is being able to draw on the behaviour of chromosomes in gamete formation. Making the connection between genetic ratios and chromosome behaviour has been shown to pose problems for American high school students.[23] And in Australia the suggestion has been made that the mechanism of inheritance may not be a suitable topic to include in the curriculum for fifteen year olds because of its difficulty![24] What are the difficulties and are they equally relevant in other countries?

Confusion over the terms 'gene' and 'allele'

Longden has pointed out that the student encounters in genetics a problem of classification known as 'inclusive seriation' where a particular object belongs to a larger class of objects, which in turn belong to a superordinate class.[25] Such a sequence is seen in allele> gene> chromosome.

His interview study of A-level biology students revealed this difficulty:

Interviewer: 'What's an allele?'
Student: 'I can't differentiate between a chromosome and an allele – there's a difference but I don't know what it is'

Longden, 1982, p. 137

We have found similar confusion with A-level students:

You get a number of genes on each . . . hang on, I'm confused now. I think two alleles make up a gene.

and

I think I'm muddled with genes and alleles. That's one thing I'm not really sure about. Sometimes we've called them genes and other times alleles.

A related misconception to that shown above is revealed in the following quotation:

Alleles are characteristics and genes are made up of different alleles. You can get several alleles on one gene.[26]

These students appear to think that a gene is made up of alleles as if the allele is a subunit of a gene. This goes beyond mere confusion of terms.

Duplication of chromosomes into chromatids in meiosis

A-level students (and some undergraduate biologists) show a persistent lack of understanding about the relationship between chromosomes and chromatids, the timing of replication of DNA in the cell cycle and the segregation of alleles in meiosis. In a recent A-level practical examination in biology[27] students were required to model the structure of a pair of homologous chromosomes at metaphase I of meiosis using pipe cleaners. They were told to use different coloured pipe cleaners (two colours were provided) to represent the different chromosomes of a homologous pair. They were then asked to label their model to show the positions of alleles if the individual was heterozygous for one gene. The correct response is shown in Fig. 7.6. How they responded to the task indicated their understanding of three main ideas:

1. Each chromosome had duplicated into identical chromatids: shown by the use of four pipe cleaners, two of each colour, tied appropriately.
2. Sister chromatids carry the same alleles: two labels carrying the same symbol attached to each pipe cleaner of the tied pair.
3. In heterozygotes, homologous chromosomes carry different alleles at the same locus: corresponding positions labelled for different alleles, e.g. AA and aa.

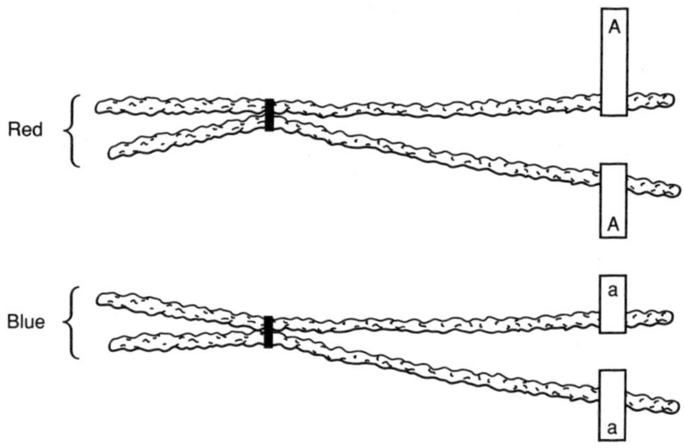

Figure 7.6 Modelling the structure of a pair of homologous chromosomes at metaphase I of meiosis using pipe cleaners

In a sample of 614 responses, 53 per cent of these A-level students appeared to have grasped idea (1) and 16 per cent both ideas (2) and (3). Many misconceptions were revealed, e.g. that sister chromatids were not genetically identical, alleles did not occupy the same locus on homologous chromosomes and that chromosome duplication had not occurred.

What was probed in this task essentially amounted to the understanding of the principles of a monohybrid cross. The respondents here were A-level students (eighteen or nineteen year olds) not the fourteen to sixteen year olds targeted by level 8c of the National Curriculum! Why do these difficulties occur? An answer may lie in the methods of teaching these topics. Often school textbooks present meiosis as a process divided into stages that seem to be convenient sub-headings under which descriptions of the events are given. Many students remember stage names at the expense of nearly everything else! Teaching needs to emphasize the dynamic nature of the process using a variety of teaching aids such as photographs of chromosomes at different stages to sort; film and video, particularly of time-lapse phase contrast microscopy of the process; and building of chromosome models such as those discussed earlier. Three-dimensional understanding is required; hence the use solely of two-dimensional images is inadequate. We should also be careful in our use of terms and how we explain them. Some like 'allele' and 'gene' must be used at all levels of work, but there are some terms that could be ignored or their use be restricted to higher levels only – 'chromatid' is such a term.[28]

In the past too much teaching of genetics has been based solely on pupils' abilities to solve pencil and paper problems such as: 'If a blue-eyed man marries a brown-eyed woman and they have four children what are the chances that . . . ', etc. A traditional starting point for many teachers has been a semi-historical treatment of Mendel's work, i.e. the ideas involved, of genotype, phenotype, alleles, their segregation, etc., have been introduced to pupils through an explanation of what Mendel did and what he inferred from his results. But Mendel was a genius and most pupils are not! They clearly must be incapable of making the same kinds of conceptual links as he did. A different approach is called for. A wide range of material is available for teaching practical genetics,[29] so it is most unfortunate that the only organism selected as an example for this purpose in the National Curriculum is *Drosophila*. This has been described as being 'one of the least suitable organisms possible for illustrating Mendel's laws'.[30]

Apart from that, for many busy teachers, the organizational demands of maintaining *Drosophila* cultures and using them to carry out breeding experiments are too great and the time-scale involved is too extensive. Other material seems to have enormous advantages, e.g. maize pollen and cobs, tomato seedlings, etc. sold as kits, from which the results of two or even three generation crosses can be presented simultaneously to pupils. However, whatever material of this sort is used, pupils have to be helped to appreciate the underlying principles, and those ideas are not made easier solely by selecting a particular organism for demonstrating them. A particularly useful procedure here is the modelling of crosses using beads.[31] This has the advantage of making the assumptions underpinning the cross explicit and tangible, e.g. a characteristic is represented by two beads, one bead constitutes a gamete, the random nature of fertilization is simulated, etc. As in the case of the pipe cleaners it is easier to talk through misunderstandings when you can, literally, pick up or point to the 'ideas' involved.

Natural selection and evolution

Although there has been some debate about the teaching of evolution as a piece of scientific orthodoxy, here I am going to confine my attention to the themes I have developed so far in this chapter. Hence these are areas that pupils have encountered outside the school and about which they will have ideas. In what can now be seen to have been a seminal study, John Deadman[32] showed that his pupils were able to express cogent opinions about a number of areas of evolution before they were taught the topic. They were aware that living organisms had changed over geological time and that characteristics were transmitted from one generation to another. Central to their thinking about these areas was the concept of adaptation. This was seen, however, as an individual response to an environmental change that had survival value, i.e. the change in the organism happened because it was needed. Essentially these are Lamarckian ideas and they run counter to the scientifically accepted ideas of natural selection. Margaret Brumby[33] has shown that these ideas may be retained, despite formal teaching, by A-level biology students who have gone on to tertiary education in the subject. Such is the current extent of television coverage of the natural history of familiar and far-away places that exposure to ideas about adaptation is even more likely now. In Brumby's study adaptation was seen as a positive process rather than as an end result of selection. She underlines the need to make clear to students the difference between 'to be adapted' and 'to adapt' and the significance of the time-scales involved: characteristics acquired during the lifetime of an organism are not inherited. Clear thinking about variation in living organisms is necessary. Students showing difficulties in this area consider that mutations are caused by environmental changes. The emphasis taken in the National Curriculum of dealing with variation in a range of living things in Key stage 3 hopefully will establish the concept appropriately for later work on the causes and implications of variations. It will be the responsibility of teachers to decide how the adaptive nature of such variations will be handled with their pupils. I hope this is done more felicitously than the example given in the National Curriculum to support the concept of competition: 'explain the *need* for woodland floor plants to grow and flower before trees bear leaves and block out the light' (my italics).

Notes and references

1. For an account of the research background to this see Bell, B. and Brook, A. (1984) *Aspects of Secondary Students' Understanding of Plant Nutrition: Full Report.* Children's Learning in Science Project, Centre for Studies in Science and Maths Education, University of Leeds. Details of the teaching approach and of appropriate practical activities are in *Approaches to Teaching Plant Nutrition*, Children's Learning in Science Project, CSSME, University of Leeds, 1987.
2. Pupils have problems in entering into the spirit of the 'game' in which they are expected to find out something that the teacher clearly knows already and that they could be told and hence learn more quickly than by doing an experiment to 'find out'. See Driver, R. (1983) *The Pupil as Scientist*? Milton Keynes: Open University Press; Wellington, J. J. (1981) What's supposed to happen, sir? Some problems with discovery learning. *School Science Review*, vol. 63, no. 222, 167–73; and Stevens, P. (1978) On the Nuffield philosophy of science. *Journal of Philosophy of Education*, vol. 12, 99–111.
3. Quoted in Bell, B. and Brook, A. (1984) op. cit. Original source is Driver, R., Child, D., Gott, R., Johnson, S., Worsley, C. and Wylie, F. (1984) *Science in Schools at Age 15: Report No 2.* Assessment of Performance Unit, London.
4. Bell, B. and Brook, A. (1984) op. cit.
5. Bell, B. and Brook, A. (1984) op. cit.
6. An approach to the teaching of plant nutrition, has been developed in New Zealand that focuses on the origin of new wood in the plant precisely because this material is more familiar to pupils than is starch. Based on constructivist teaching strategies the authors claim that the teaching approach results in significant improvements in pupils' learning. See Barker, M. and Carr, M. (1989), Photosynthesis – can our pupils see the wood for the trees. *Journal of Biological Education*, vol. 23, no. 1, 41–4.
7. Thirty-one such ideas have been listed by Wandersee, J. H. (1983) Students' misconceptions about photosynthesis, in Helm, H. and Novak, J.D. (eds) *Proceedings of the International Seminar: Misconceptions in Science and Mathematics*, 441– 66. Ithaca: Cornell University. Examples of teaching approaches that start with the presence or absence of starch in leaves abound – you will probably remember them from your own studies or even see them written up in existing school schemes of work. For documented examples see Bell, B. (1985) *The Construction of Meaning and Conceptual Change in Classroom Settings: Case Studies on Plant Nutrition*, Children's Learning in Science Project, CSSME, University of Leeds.
8. Simpson, M. and Arnold, B. (1982) Availability of prerequisite concepts for learning biology at certificate level. *Journal of Biological Education*, vol. 16, no. 1, 65–72. Pupils have difficulties in handling a range of important ideas about the particulate nature of matter that are relevant here

and in a large number of other areas in biology. For a consideration of these difficulties see Brook, A., Briggs, H. and Driver, R. (1984) *Aspects of Children's Understanding of the Particulate Nature of Matter*. Children's Learning in Science Project, CSSME, University of Leeds.

9. For example, Johnstone, A.H. and Mahmoud, N. A. (1980) Pupils' problems with water potential. *Journal of Biological Education*, vol. 14, no. 4, 325–8.
10. Details of the approach and an evaluation of it are in Johnstone, A.H. and Mahmoud, N. A. (1981) Pupils' responses to a model of water transport. *Journal of Biological Education*, vol. 15, no. 3, 203–8. For an example of a GCSE textbook which utilizes the ideas see Mackean, D. G. (1986) *GCSE Biology*. London: John Murray.
11. Piaget, J. (1929) *The Child's Conception of the World*. New York: Harcourt Press.
12. Tamir, P., Gal-Choppin, R. and Nussinovitz, R. (1981) How do intermediate and high school students conceptualise living and non-living? *Journal of Research in Science Teaching*, vol. 18, no. 3, 241–8.
13. Simpson, M. and Arnold, B. (1982) op. cit.
14. See in particular chapter 3 in Osborne, R. and Freyberg, P. (1985) *Learning in Science. The Implications of Children's Science*. London: Heinemann.
15. Or what the National Curriculum for Science calls 'life processes'.
16. Deciding when this might be presents some problems if you teach in England and Wales; see the Statements of Attainment in the National Curriculum for Sc2 strand (i) at levels 3, 5 and 7! My own view is that these are year 9 and year 10 topics.
17. This work is reported in Borrell, P. (1987) *A school-based study to investigate secondary pupils' understanding of the concept respiration*, M.Ed. dissertation, University of Hull, 1987.
18. Shayer, M and Adey, P. (1981) *Towards a Science of Science Teaching*, p. 3. London: Heinemann Educational Books.
19. Ryman, D. (1974) Children's understanding of the classification of living organisms. *Journal of Biological Education*, vol. 8, no. 3, 140–4.
20. See Osborne, R. and Freyberg, P. (1985) op. cit.; Bell, B. F. (1981) When is an animal not an animal? *Journal of Biological Education*, vol. 15, no. 3, 213–18; Bell, B. and Barker, M. (1982) Towards a scientific concept of 'animal'. *Journal of Biological Education*, vol. 16, no. 3, 197–200.
21. For details of such methods see Adey, P. S., Shayer, M. and Yates, C. (1989) *Thinking Science: Student and Teacher's Materials for the CASE Intervention*. London: Macmillan.
22. Longden, B. (1982) Genetics – are there inherent learning difficulties? *Journal of Biological Education*, vol. 16, no. 2, 137.
23. Stewart, J. (1983) Student problem solving in high school genetics. *Science Education*, vol. 67, no. 4, 523–40.
24. Hackling, M.W. and Treagust, D. (1984) Research data necessary for meaningful review of grade ten high school genetics curricula. *Journal of Research in Science Teaching*, vol. 21, no. 2, 197–209.

25. Longden, B. (1982) op. cit.
26. Quotations from a study for a B.Sc. dissertation in the University of Hull by Robert Desborough (1992).
27. Set by University of Cambridge Local Examinations Syndicate in 1989, full details of the analysis are in Brown, C. R. (1990) Some misconceptions in meiosis shown by students responding to an Advanced level practical examination question in biology. *Journal of Biological Education*, vol. 24, no. 3, 182–6.
28. Guidance about the use of appropriate terminology in biology is available in *Biological Nomenclature, Recommendations on Terms, Units and Symbols*, Institute of Biology, 1989.
29. See the catalogues of the major biological supplies agencies.
30. Radford, A. and Bird-Stewart, J. A. (1982) Teaching genetics in schools. *Journal of Biological Education*, vol. 16, no. 3, 179.
31. Details are available in *Revised Nuffield Biology, Text 4. The perpetuation of life*. London: Longman.
32. Deadman, J. A. and Kelly, P. J. (1978) What do secondary school boys understand about evolution and heredity before they are taught the topics? *Journal of Biological Education*, vol. 12, no. 1, 7–15.
33. Brumby, M. (1979) Problems in learning the concept of natural selection. *Journal of Biological Education*, vol. 13, no. 2, 119–22.

CHAPTER 8

Assessment in biology

Introduction

Confusion sometimes arises over the use of the two words 'assessment' and 'evaluation'. Indeed in some parts of the world, notably North America, they are used synonymously. In Britain, 'assessment' involves finding out something about pupils, and 'evaluation' gaining information about the quality, utility, etc. of lessons, textbooks, schemes of work or whole curricula. It is in this way that assessment will be treated here.

Some years ago teachers taught and examination boards examined. Assessment of pupils' work was, of course, carried out by teachers, but such assessments were of an informal nature, directed towards detecting deficiencies in learning and such information remained confined within the classroom. However, with the development of course work assessment of various types in GCSE and A-level syllabuses, and particularly with the implementation of the National Curriculum, teachers' assessments have taken on greater significance in contributing to formal, external reporting of pupils' performances. The wide spectrum of areas involving assessment in which you, as a teacher, can expect to be involved includes:

1. setting and marking routine classwork and homework tasks;
2. setting, administering and marking end-of-topic, end-of module, termly or yearly tests;
3. making judgements about progress from classroom interactions involving listening to pupils' talk, watching them conduct activities such as those in practical work;
4. assessing and recording for National Curriculum reporting purposes and for reporting to parents;
5. helping pupils compile their records of achievement;
6. marking course-work assignments for GCSE and A-level examinations;
7. assessing practical skills, e.g. for A level.

That there are difficulties presented in these fields is indicated by the following comments from the Secondary Science Curriculum Review:[1]

> in our experience, good assessment techniques . . . are found in few schools. This is particularly true of pen and paper tests of any sort. We would argue that there are essentially three reasons for this problem:
>
> 1. Few teachers are clear enough about the real purpose of the assessment.
> 2. Construction of assessments has to take place in a very busy calendar
> 3. Whilst all teachers are trained to teach at least one area of the curriculum, few are trained in any adequate sense to assess children's progress in it.

It is likely that these difficulties may have increased as a result of the requirements involving National Curriculum assessment in recent years! Hence the learning curve that faces you in this area is a very steep one. You will be able to draw on the expertise of your new colleagues and you will be operating within the framework of the school's assessment, recording and reporting policy. Nevertheless one of the most difficult aspects of your role is in piecing together your own ideas about standards of work. And you will have to wrestle with such difficulties very early in your teaching career: pupils and their parents will expect to receive an accurate record of progress to which some predictive weight may be attached. For example, 'Jenny is doing quite well at her GCSE Science (Double Award), should she be thinking about doing A level biology, she is toying with the thought of reading medicine?' Or, 'John is interested in horses and wants to apply to veterinary school, what do you advise?' Your judgements in these situations will depend on the information that you have gained about your students as a result of the various assessment procedures you have used. These will have given you some feedback about their attainment. Equally important are the processes by which we give our pupils feedback that will support their continued learning. Such information is essentially formative rather than summative. Managing your teaching to obtain such information and finding opportunities to use it fruitfully are issues of considerable current importance in science teaching.

Some principles of assessment

From the list of assessment-related activities given earlier you will see that two main purposes are involved: providing information for selective reasons and for diagnostic ones. It is important to be clear about the function of any proposed piece of assessment: quite different techniques may be required to provide the class teacher with the information that a particular pupil is ready to move on to the next activity from that needed to present to a potential employer or higher education course selector. These techniques may also differ sharply as far as the kind of information they generate is concerned. However, there are particular issues that underpin all assessments, irrespective of nature and purpose. These arise because assessment has essentially to do with *measuring* something. These are discussed below; later in the chapter they will be re-examined for their importance in the role of the class teacher.

Reliability

If we are interested in measuring the length of an object we can use a ruler. How effective would our measurement be if the ruler was not evenly marked out in units such as centimetres and millimetres? And if two different people were using a correctly calibrated ruler to make the same measurement would they come up with the same value? With a real ruler, of course, the measurement should be the same irrespective of the measurer and on every occasion when the measurement is made. Hence the measurement would be totally reliable. So should be our assessment 'rulers'. Hence reliability can be thought of in terms of the precision (accuracy) and the reproducibility of such measurement. It depends on some aspect of the measuring instrument and on some features of the performance of the measurers. So your written test would be completely reliable if, when you gave the test to a group of students on two successive days, the rank orders produced on the two occasions were identical. If you did this (you are unlikely to do so for fairly obvious reasons) you would find that your test was not totally reliable, no test is. Even though it is not usually possible (or sensible) to repeat tests in this way, there are several other methods by which the reliability of a test can be estimated. The most extensively used are the so-called *internal consistency methods*, which involve correlating scores obtained on half the questions constituting the test with those obtained on the other half.[2] On this basis reliability is measured in terms of the correlation coefficient between scores and thus it ranges from +1 to –1.

Robert Ebel[3] gives a useful example of the significance of differences in the magnitude of the reliability coefficient of a test. If a group of examinees is divided into upper and lower halves on the basis of a test and the test is repeated on the same candidates, the changes in membership of the two groups are related to test reliability as follows:

Reliability of test	1.00	0.96	0.90	0.85	0.81	0.76	0.64	0.49	0.25
Percentage of candidates staying put	100	95	90	87	85	83	80	74	66

As a busy class teacher you will probably not have the time to calculate reliability coefficients for your written tests and examinations, so you will be working on the basis that your tests are reasonably reliable and not too many injustices are being done to your pupils! Research studies have produced some useful information about the reliabilities of various examinations and there are some straightforward principles derived from them that can be used as guidelines for teachers interested in maximizing the reliabilities of their tests. These are summarized below:

1. Longer tests are more reliable than shorter ones, so within reason (bearing in mind likely attention span and fatigue factors) use longer tests.
2. Allow adequate time for the pupils to complete the test.
3. Fixed response tests (including multiple choice) and short answer question tests tend to be more reliable than essay and other extended response tests (also see later).
4. Each concept or principle that you wish to assess should be assessed by several questions not just one.
5. Do not include questions with different formats in the same section of a test, e.g. multiple choice questions should appear *en bloc* rather than being scattered among questions of a different format.
6. Avoid giving a choice between questions or reduce the element of choice as far as possible.
7. Avoid ambiguous instructions in the presentation of the task. Ensure that your students understand the different responses expected by instructions such as ‘discuss’, ‘describe’, ‘compare’, etc.
8. Eliminate the possibility of marking variability by marking to clear, unambiguous mark schemes. There is, of course, no marker unreliability in objective questions, so called because they can be *marked* objectively.

Validity

A golfer tees up the ball on a short hole, addresses the ball, swings . . . and the ball disappears out of bounds. No recourse but to try again! Exactly the same thing happens and when the golfer eventually goes to retrieve the two balls they are found lying side by side! The golfer’s swing has resulted in the ball following the same trajectory on the two occasions; if it did so every time, the swing would be totally reliable. But clearly in this case reliability is insufficient; the swing must also result in fulfilment of the purpose of the activity as completely as possible. A totally *valid* swing would result in a hole-in-one (I said it was a short hole). Of course if the swing was totally reliable as well, a hole-in-one would be achieved every time.

In assessment terminology the commonest meaning attributed to 'validity' of a test is given in the following much-quoted definition:

> The accuracy with which it measures that which it is intended to measure or the degree to which it approaches infallibility in measuring what it purports to measure.[4]

In practice this most frequently involves analysing the extent to which a test or examination samples the content and/or the range of objectives of the course or curriculum to which it relates. This is the so-called content validity of a test. Examining boards attempt to guarantee the validity of their examinations through the use of specification grids. In these, assessment objectives and content areas of the syllabus are defined. The relative emphasis given to these in examinations is specified in terms of mark weightings. Relative mark weightings reflect the time devoted to teaching the various components of the matrix. An example is shown in Table 8.1.

Table 8.1 **Example of a specification grid**

	Assessment objectives		
Content	A. Knowledge with understanding	B. Handling ideas and information C. Practical activities	Mark allocations in complete assessment
Theme I. Diversity of organisms			Between 5 and 10%
Theme II. Relationships between organisms and with the Environment			Between 25 and 40%
Theme III. Organization and maintenance of the individual			Between 25 and 40%
Theme IV. Development of organisms and the continuity of life			Between 15 and 25%
Mark allocation in complete assessment	Of the order of 45%	Not less than 40% (at least 20% to experimental and observational skills)	

Hence it is possible to analyse the extent of the match between the specification grid and the mark weightings used in the examination papers, provided it is possible to assign the questions to particular content areas and skills or processes. In public examinations extensive mismatches between paper weightings and specification grid are rarely found. Appropriate analyses are usually conducted on papers in the assembly stage before they are put into operation with candidates.

For the class teacher the key to developing and using valid assessment procedures lies in the clear identification of objectives, particularly if they have been framed as observable learning outcomes. Then what has to be assessed is obvious; the task is to use techniques that are appropriate for the particular purpose involved. This point is developed throughout the remainder of this chapter.

Criterion-referenced and norm-referenced assessment

The most significant difference between criterion-referencing (CR) and norm-referencing lies in how the results of the assessment are *reported.* Thus norm-referencing occurs when information about pupils' attainment is related in some way (usually by grades) to that of their peers. CR is based on gaining information about pupils' achievement in relation to specified curriculum objectives. A fuller definition of CR assessment is as follows:

> Assessment that provides information about the specific knowledge and abilities of pupils through their performances on various kinds of tasks that are interpretable in terms of what pupils know or can do, without reference to the performance of others.[5]

It may be that you are experiencing the feeling that these distinctions are fairly esoteric and seemingly distant from the world of the classroom teacher. Not so! Over the last few years we have seen a powerful movement towards the use of CR in both GCSE and A-level assessment, and the strengths, weaknesses and practicability of conducting CR assessment lies at the heart of the debate about testing in the National Curriculum. Such issues will immediately engage you when you begin teaching.

As we said in norm-referencing pupils are placed in categories on the basis of their performances. The percentage distributions of the categories will be predetermined. For example, in an end of year examination we may decide that the grade distribution will be as follows:

Grade A: gained by top 10 per cent of pupils
Grade B: by the next 15 per cent
Grade C: next 50 per cent
Grade D: next 15 per cent
Grade E: next 10 per cent

For placement in these grades to be as error-free as possible one would want to ensure that the assessment procedures used showed a high reliability and that the marks of the pupils were spread over a wide range. Even so, we may be drawing grade boundaries between very similar levels of performance. The result of achieving a wide range of marks is obvious: some of the pupils will score low marks. This can be a profoundly demotivating experience if it happens to the same pupils in most subjects in the curriculum. You might think about the purpose of such a grading system and the usefulness of the information generated. It is fairly obvious that it has little formative use (by which difficulties in learning can be detected and addressed subsequently by teacher and pupil), although there will be potentially useful information buried in the pupils' responses to the test questions.

Issues of this sort have led to the upsurge of interest in CR assessment, and its 'can-do' style of reporting.[6] Figure 8.1 is an example of a test item from a CR system.[7] The example illustrates one of the key requirements of CR assessment – that the task tests one specific outcome, an objective of the course of study. Pupils' responses fall into two possible categories: correct or incorrect. However, there are a number of issues here that, as we will see, render the apparent precision of CR assessment open to question. We could not generalize from this one result to be able to say that the pupil is able to take accurate readings from the scales of measuring instruments. To feel confident about making such a judgement you would want to look at

Instruction to pupil

Your task is to read the ammeter scale and write down your answer.

Be as accurate as you can.

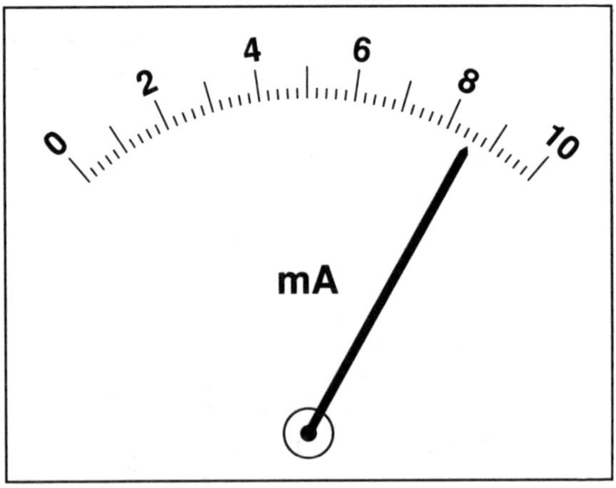

Figure 8.1 Example of a test item from a criterion-referenced system

performance on a variety of different tasks, e.g. an ammeter scale, a meniscus level in a measuring cylinder, a capillary tube or a thermometer. If this were to be done, two further issues would arise. Would you expect each measurement to be accurate or are there acceptable limits of error? And in order to qualify for a 'can-do', what level of performance is expected across the range of tasks, e.g. all correct, nine out of ten correct and so on? A further problem arises from the need for each test question to test a single objective – it would appear to be necessary to construct very large numbers of test questions to be confident that representative sampling of curriculum objectives had occurred. Then the problem is how to aggregate this information for sensible reporting.

Two examples will serve to highlight some of the problems facing teachers. A number of A-level boards have systems of internal assessment of practical skills in which teachers assess their students' performance using guidelines supplied by the board. One A-level board identifies three such skills, each of which is assessed on a scale of 1–6. One such skill is described as follows:

Observing, measuring and recording

The candidate:

2/1 experiences considerable difficulty in reading scales and in making observations from the microscope; fails to record results systematically; produces graphs, tables and drawings that are incomplete or incorrect.

4/3 takes readings and makes observations from the microscope that are adequate but may be incomplete; records results but may need some help in selecting the correct format for presentation.

6/5 reads scales and observes microscopic specimens accurately; records the results in a suitable systematic form; produces graphs, tables and drawings that are complete and correctly labelled.

Two main problems arise in putting this into operation. First, the 'skill' involves a whole range of tasks: reading scales, observing microscope specimens accurately, determining the form in which results will be recorded, producing graphs, etc. that are complete and correctly labelled. Despite this a single numerical value has to be identified for the skill. Either the students are expected to be equally good (or bad) at reading scales as they are at observing specimens, etc. or each of the sub-skills must be considered separately and then the results aggregated in some (unspecified) way. Another difficulty is the subjectivity of the decision-making about the scale points that are not given written descriptors. What are the different levels of performance that would gain a 5 as opposed to a 6?

The second example is drawn from the 1991 National Curriculum Science order. National Curriculum assessment purports to be CR. Selecting a statement of attainment (almost at random) we read in Sc2, 5a:

> [pupils should] be able to name and outline the functions of the major organs and organ systems in mammals and flowering plants.

From a teaching perspective, the immediate question is: what *are* the major organs, etc. that must be dealt with so that pupils' knowledge can be tested? Unfortunately no clarification of this exists. Currently, the 'slimming down' of the National Curriculum proposed by Dearing is taking place presumably resulting in deletion or amalgamation of existing SoAs. The net result of this must be that, in the future, even less guidance will exist than at present. An example is given in Fig. 8.2 of a test question that addresses the descriptor of Sc2, 5a. As you see it clears up some of the questions about detail that one could want clarified, but it leaves others unanswered. It also illustrates the difficulty facing the test constructor – how many such questions covering what examples of 'major organs' is it necessary for candidates to answer before we can say, with confidence, that they have reached level 5?

The diagram shows a section through a flower.

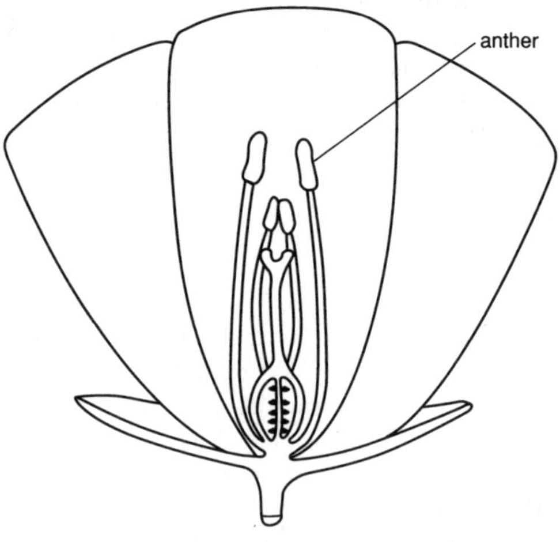

(i) On the diagram, label and name the **three** female parts.

(ii) What is the job of the anther?

Figure 8.2 Example of a test question that addresses the descriptor of Sc2, 5a

Diagnostic assessment

At the heart of this is the idea that assessment can be used to identify learning difficulties and this can lead to structuring a programme of appropriate remedial measures. That being so, diagnosis should go deeper than the identification of errors; it should probe the underlying causes for failure. It will address itself to exposing some of the preconceptions, misconceptions, etc. dealt with in the previous chapter. Wood makes the point that 'Trying to do diagnostic work with tests as currently conceived is almost certainly a waste of time.'[8] This is because the conventional tests do not fulfil the requirement stated above; they concentrate on identifying errors (sometimes) but do not go beyond that. A Scottish study has concluded that diagnosis of learning difficulties should involve the following questions:

What does the pupil 'know' when a particular wrong answer is given to a test question?
Why does that particular answer make more sense to the pupil than the correct answer?
What does the pupil need to know to answer the question correctly?
Did the learning experience offered in class make it clear to the pupil what was to be learned?
Was the information given unambiguous, were the critical ideas salient, and were there inherent possibilities of confusion?
What did the pupil need to know to understand the taught material?
Was that prior knowledge made available in previous courses?

Discussion between the teacher and pupil was found to be particularly fruitful in this study:

> These discussions provided much clearer clues to the sources of learning difficulties than were given by class tests . . . and we were left in no doubt that listening to pupils talk should be regarded as the most important single diagnostic activity of the teacher. It is an illusion to think that pupil learning difficulties can be detected and defined by 'diagnostic tests' alone and it is inevitable that difficulties will continue to remain undiagnosed until listening to pupils plays a much larger role in classroom activities.[9]

It is difficult to find time in the busy classroom to conduct in-depth discussions, however desirable they may be; hence written methods of achieving the same purpose are particularly useful. Such methods are different from conventional tests and are not extensively used in the UK. Three examples are given below.

Two-tier multiple choice tests

Here is an example of a question from a test on photosynthesis and respiration developed and used in Australia.[10]

Which gas is taken in by green plants in large amounts when there is no light energy at all?

(1) carbon dioxide gas
(2) oxygen gas

The reason for my answer is because:

(a) This gas is used in photosynthesis which occurs in green plants all the time.
(b) This gas is used in photosynthesis which occurs in green plants when there is no light energy at all.
(c) This gas is used in respiration which only occurs in green plants when there is no light energy to photosynthesize.
(d) This gas is used in respiration which takes place continuously in green plants.
(e) ______________________________

The second part of the test item has been prepared from common misconceptions that have been identified from groups of pupils. Although the development of tests of this sort is time-consuming, such tests have a number of uses:

1. before teaching a topic in order to map the ideas that students bring to the situation;
2. to structure teaching approaches appropriate for dealing with individual preconceptions;
3. after teaching the topic, responses to the questions can be used for group discussion enabling reflection on their own reasoning and that of their peers.

Some sample results for this item are shown in Table 8.2 for students who had been taught the topics. As you can see, there is plenty of scope for discussion.

Table 8.2 **Results of a two-tier multiple choice test on photosynthesis and respiration**

Year level	Number of students	Content choice	Reason choice (% of students) a	b	c	d	e	No reason
8	137	1	14.6	16.8	10.9	16.1	–	2.9
		2	7.3	10.2	11.7	8.8*	–	0.7
9	88	1	11.4	5.7	20.5	17.0	–	1.1
		2	3.4	18.2	11.4	11.4*	–	–
10	99	1	8.1	11.1	11.2	24.2	4.0	–
		2	3.0	8.1	23.3	7.1*	–	–

* The correct choice and reason.
– No response.

Bur diagrams

These are diagrams by which we can explore the meanings of, and connections between, words and concepts. The name comes from the sticky burs that are the fruits of burdock. Just as these will stick by their hooks to objects, so the words shown inside the bur will stick to associations in the minds of learners. The greater the number of hooks that can be labelled, the more elaborate is the concept held by the learner. Sutton gives the following examples for the concept 'fruit' (Fig. 8.3).[11] The first diagram shows how the concept is associated with a range of everyday connections. The second shows a greater elaboration of the concept and it includes aspects of the scientific meaning that, conventionally, are shown by bold hooks. Figure 8.3 illustrates how the concept might have been extended as a result of some teaching and hence provides an indication of the learning that has taken place. Bur diagrams are also fun to complete and provide another possible context by which discussion about meanings of words and ideas can be developed with the pupil. They are also excellent devices for revision activities for pupils as well as summaries of units of work.

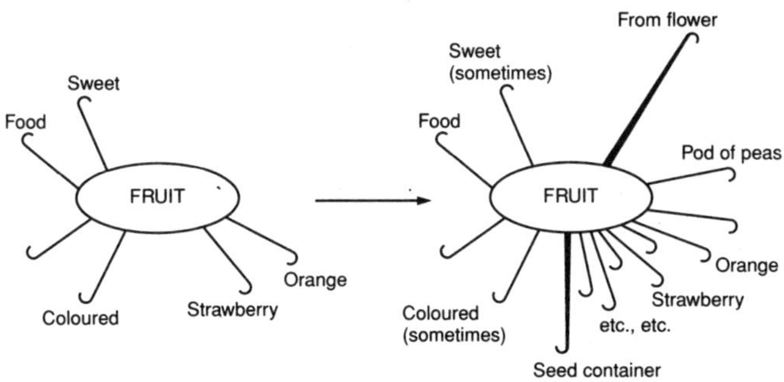

Figure 8.3 Bur diagrams for concept 'fruit'

Concept mapping

This is another method of exploring the range of ideas held by students about scientific concepts. Beginning as a research tool, concept mapping has found its way into relatively extensive classroom use, but not in the UK. In the USA, pupils of primary school age have been shown to be able to draw up concept maps. For older students, at least one biology textbook[12] represents the knowledge areas and their interrelationships in each chapter by means of concept maps. Several different research studies have shown that students' learning can be made more effective through the consistent use of concept-mapping

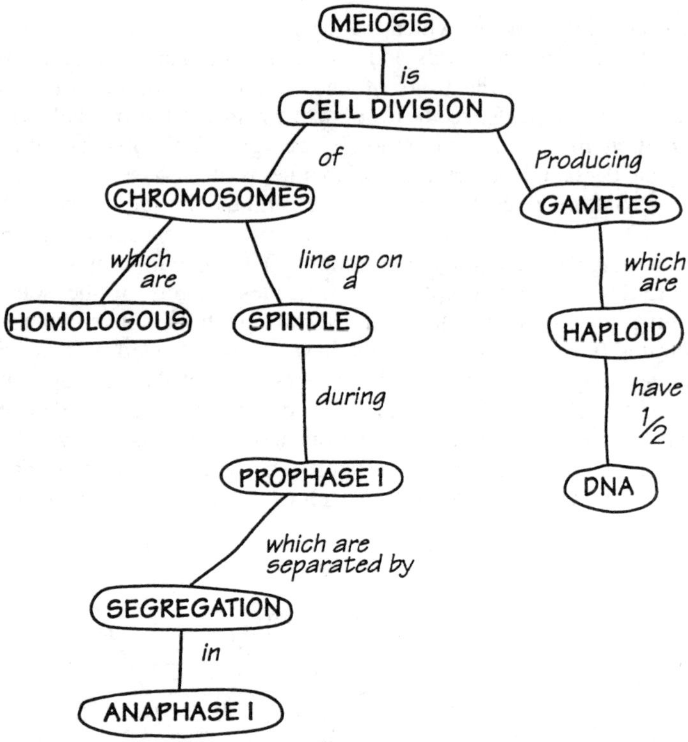

Figure 8.4 Example of a concept map of 'meiosis'

techniques. More particularly here, they also provide a way of detecting misconceptions and gaps in understanding of concepts. Figure 8.4 shows a map drawn by an A-level student about the concept 'meiosis'.

The structure of such maps is not haphazard but follows a set of rules:

1. The concepts are ordered from greater to lesser degrees of generality, i.e. from superordinate to subordinate concepts. The more generally inclusive concepts are written at the top of the page, leading down to the more specific ones (Fig. 8.5). This arrangement thus produces a hierarchy of levels (four in this case).
2. The concepts are joined by a connecting line and labelled with a linking word that the learner sees as the connection between the concepts. Connecting lines can also be drawn between concepts at the same level in the hierarchy.

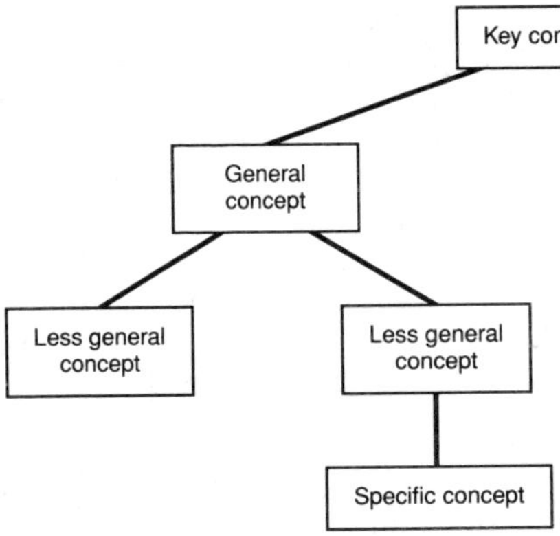

Figure 8.5 Structure of a concept map

Novak, who developed the technique of concept mapping, advocates a two-step approach to introducing the procedure to students. A series of preparatory activities introduces pupils to the ideas of concepts and linking words. These include writing short sentences to identify concept words and analysing appropriate passages of text to identify key concepts, less important concepts and linking words. This is followed by a structured approach to concept mapping. The following is an amended version of Novak's steps for concept-mapping activities.[13]

1. Select a particularly meaningful paragraph or two from a text or other printed material. Have the students read the text and select the key concepts, i.e. those concepts necessary for understanding the meaning of the text. List these concepts on the board (or overhead projector) as they are identified. Now discuss with the students which concept is the most important, most inclusive idea in the text.
2. Put the most inclusive concept at the head of a new list of rank-ordered concepts. List the next most general, most inclusive concepts, working through the first list until all concepts are rank ordered. There will not always be agreement among the students on the ordering, but usually only a few major differences in ranking of the concepts will arise.
3. Now begin constructing a concept map, using the rank-ordered list as a guide in building the concept hierarchy. Have students help in choosing good linking words to form the propositions shown by the lines on the map. One good way to have them practise map making is to have students write concept words and linking words

on paper rectangles and then rearrange these rectangles as they get new insights on the map organization.

4. Now look for cross-links between concepts in one section of the map and concepts in another part of the concept 'tree'. Have students help to choose linking words for the cross-links.
5. Most first effort maps have poor symmetry or some concept clusters poorly located relative to other more closely related concepts or clusters of concepts. Reconstruct the map if this would be helpful. Point out to students that at least one and sometimes two or three reconstructions of a map are needed to show a good representation of propositional meanings as they understand them.

In the procedure described above, concepts are derived by presenting the students with some text. Alternatively, students can be supplied with lists of the concepts that are to be mapped, presented in a random fashion rather than in a hierarchy. A further possibility is to provide a key word, e.g. 'food', and have the students generate the concepts that they associate with this as in the case of bur diagrams. Clearly these procedures place slightly different demands on the respondents.

Concept maps can be scored numerically and show adequate reliability; hence they could be incorporated into written achievement tests. A number of different scoring procedures have been advocated, but all attempt to reflect the complexity of the maps in an appropriate way. Here is one scoring method:[14]

1. Branching: a score of 1 is given for every concept made that has two or more branches coming from it.
2. General to specific: if a concept is more general than those branching from it, a general-to-specific trend exists. A score is derived as follows:

Per cent concepts showing general-to-specific pattern	Score
0–10	0
11–29	1
30–49	2
50–69	3
70–89	4
90–100	5

3. Hierarchy: 1 point is given for a correct relationship between successive levels. That is, each subordinate should be more specific that the concept above it.

4. Technical terminology: the number of correctly used terms is counted and their percentage of the total number of terms used in the map is calculated. If the list of terms has been given to the

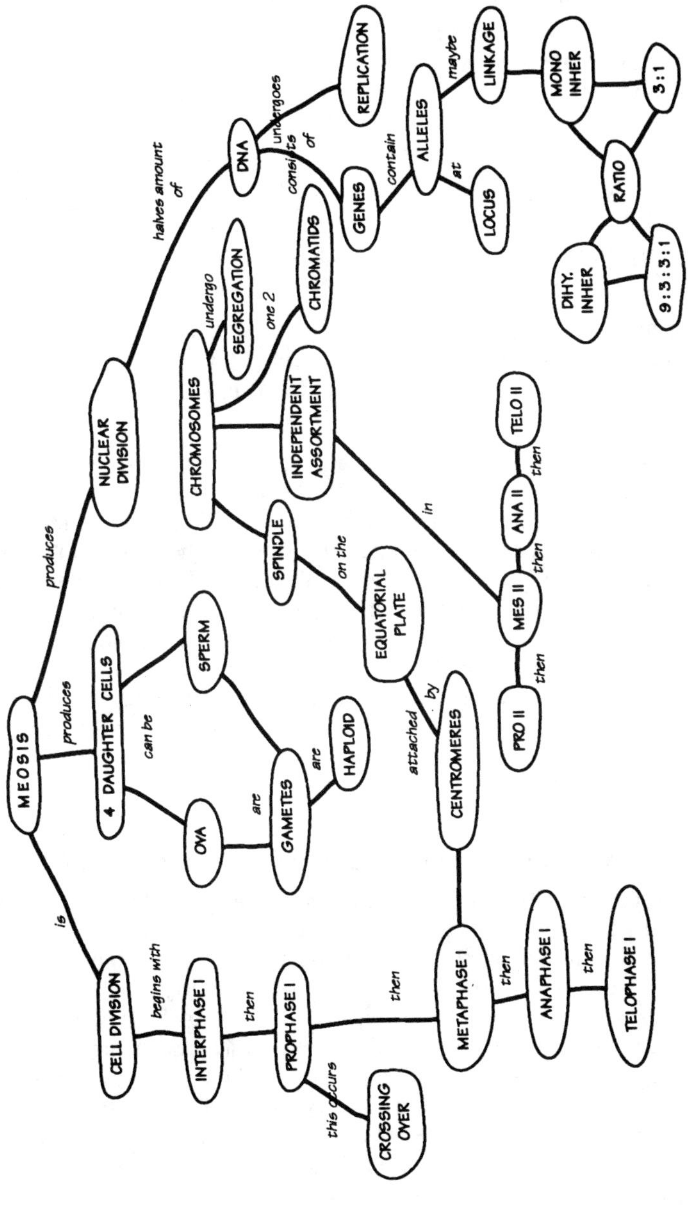

Figure 8.6 Another example of a concept map of 'meiosis'

students the percentage should be relative to that total number. Percentage is converted into a score as in (2).

5. Linking words: 1 point is given for each different valid connection made (by line) and labelled.
6. Cross-links: these show connections between different segments of the map and indicate creative or integrative thinking. 1 point for each shown.

The map for the concept 'meiosis' illustrated in Fig. 8.4 was derived by presenting the student with a list of forty-one terms that might have been used to produce the map. Figure 8.6 shows the map of a different student responding to the same task. Using the method detailed above, the map in Fig. 8.4 is scored as follows:

Branching 2	Technical terminology 1
General to specific 3	Linking words 8
Hierarchy 3	Cross links 0
	Total 16

The differences between the two responses are striking even at a qualitative level. Figure 8.6 indicates that, in the mind of the learner, the process of meiosis is effectively linked with genetic events, for example in explaining ratios. This seems not to be the case in Fig. 8.4, and hence the map provides powerful diagnostic information for the teacher and an obvious point of intervention for some remedial work. Novak and Gowin recommend that the criteria for scoring concept maps are explained to students who are then invited to discuss each others' maps, to score them, and subsequently review with the teacher how their maps might be amended to produce higher scores. You might like to conclude this section of the chapter by trying out the scoring procedure on Fig. 8.6.

Some common techniques used in written assessment

You will be familiar with the wide range of procedures used for making written assessments in biology of cognitive skills such as recall of facts, making deductions from data, analysing experimental findings, etc. These procedures range from multiple choice tests through to project reports. They are used extensively in school by teachers and by examination boards. Since these procedures are well known and further sources of information about them are readily accessible,[15] I will concentrate on identifying some aspects of their use that are of significance. The strengths and weaknesses of these procedures are summarized in Table 8.3. Below, further comment is made on some of the points summarized in this table. One general point is that the more extensive the range of procedures you use, the more likely you are to obtain information that will allow you to build up a valid picture of the achievement of your students.

Table 8.3 **Advantages and disadvantages of different assessment procedures**

Assessment method	Advantages	Disadvantages	Key points for classroom teacher
Multiple choice (and other objective types)	Test a wide range of objectives in a short time. Marked objectively. Highly reliable	Cannot be used to test written communication, synthesis of ideas, etc. Provide limited amount of feedback	Boys outperform girls on multiple choice questions, so beware gender bias. Do not attempt to set them yourself. Set as a battery not as part of a mixed format test
Short answer and structured questions	Test a wide range of objectives. Reliable. Can be used to explore different but related ideas by development through the question. Good teaching material. Gender fair. Easy to mark	Element of subjectivity in deciding correctness of response. Longer response time leads to less efficient syllabus coverage than multiple choice	Can be written by teachers. Should be an increasing level of difficulty through the question. Answer to one part should not depend on correct response to previous part. Beware making them too difficult. Do not give choice between them
Comprehension, questions on a passage, data response	Test a variety of skills including interpretation and synthesis. Good teaching material	Long response time results in even less efficient syllabus coverage. Marking can be subjective	Do not offer choice between or within questions – reading time is too great. Difficult to find appropriate and interesting material. Promising material, e.g. newspaper articles, needs extensive re-drafting. Produce a point-by-point marking scheme
Essays (extended response)	Give opportunity for synthesis of ideas and other communication skills. Allow show of originality. May reveal errors of understanding	Coverage of syllabus is poor. Lower reliability. Candidates in exams notoriously bad choosers. Difficult to set essays of similar demands. Time-consuming to mark	Reduce choice to a minimum. An effective mark scheme is essential. Beware gender bias – girls are favoured by this mode of response
Projects	Provide opportunity for a wide range of skills to be assessed. Motivating for pupils	Outcome is uncertain. Pupils can get bogged down. Greedy on time, space and equipment. Generation of ideas is demanding	Need to mark to a structure, e.g. a system of operational divisions (available from some exam boards). Mainly A-level use

Gender bias

In biology we are less plagued by issues of gender bias than in the physical sciences: the content of biology is not seen to be demotivating by boys to any greater extent than by girls. This is not quite the same thing, however, as commenting on issues of subject selection in the curriculum. The tendency for different patterns of selection to occur for physics and biology by boys and girls is well documented. Research studies have indicated that gender bias may also occur in assessment in two different ways:

1. Particular *formats* of tests may differentially favour boys or girls.
2. The particular *context* in which a question is couched may appeal, or be more accessible, to boys or girls.

Hence boys tend to outperform girls in a number of different curriculum subjects on multiple choice tests purely because of the test format. Conversely, girls are favoured by extended response questions. There is some evidence that short answer questions and structured questions are gender neutral. Multiple choice tests and essays are useful, despite their potential gender bias, for the range of reasons given in the table. The alert teacher will ensure that a balance is attained in the types of questions to which pupils respond.

The issue of the context or content of questions has been analysed by Patricia Murphy drawing on APU testing procedures:

> questions which involve such content as health, reproduction, nutrition and domestic situations are generally answered by more girls than boys. The girls also tend to achieve higher scores on these questions. In tasks which have a more overtly 'masculine' content e.g. building sites, race tracks, information from spare part catalogues, or anything with an electrical content, the reverse is true.[16]

Such effects are very subtle but they may be sufficiently marked to override the potential bias of format discussed above. Your responsibility as a class teacher is to be on the look-out for potential situations of this kind and to avoid them. Exactly the same principle applies to the need to counter potential racial or cultural bias in assessment procedures.[17]

Sources of assessment material

Where do you get the various questions that you need to use? Luckily, vast collections of material are available for your examination classes from the past papers and syllabus support material from examination boards. My strong advice is that you use these, perhaps with some judicious modification to suit your precise needs, but that you do *not* attempt to set the bulk of these questions yourself, at least not in the first stages of your career. Setting effective questions (of all formats) demands time, expertise and particularly training. The questions set by examination boards, whilst not being beyond criticism, will have undergone extensive trialling and moderating procedures. In the case of multiple choice questions the trialling will have involved analysing the level of difficulty (facility) of each question and the efficiency with which it discriminates between highest and lowest achievers. Indeed, selection for use of these questions is based on such information, which is also available to you because it forms part of each examination board's published annual reports. Similarly, examination boards publish mark schemes for all written papers including information about the assessment of projects or individual assignments. Such information is obviously so useful that it is impossible to ignore and you will quickly find yourself putting it to good use for the benefit of your pupils.

Assessing practical skills and processes

Considerable debate has occurred in this area not only in terms of the place of skills and processes in science teaching (some of these issues were touched on in Chapter 3) but also about how they should be assessed. One of the problems lies in the difficulty we have in defining skills and processes. In 1984, the Secondary Science Curriculum Review team defined a skill as 'a specific activity which a student can be trained to do', while a process was 'a rational activity involving the application of a range of skills.'[18]

In 1987 they avoided any definitions because 'almost every SSCR group using the terms has used them in a different way'.[19] Fairbrother suggests that the terminology argument should be by-passed by simply deciding that there are certain things that we want students to do; he calls them 'ings'.[20] These are all actions of various sorts, e.g. observing and measuring. Some can be observed taking place and could be assessed at that point if we wished. Others, such as thinking, cannot be assessed directly; performance must be inferred from indirect evidence like the quality of oral or written responses. We do not have a free hand in deciding what it is we want students to do, or indeed how this will be assessed. At least in part, this is determined for us by the Government (for the National Curriculum) and by GCSE and A-level boards. Details of these requirements follow in this chapter.

For assessment purposes, we will find ourselves dealing on occasions with laboratory-based procedures and on others through more conventional written tests. This variation in the test format needed to assess different science process skills is shown in the framework developed by the APU (Table 8.4).

Test materials developed by the APU, after extensive piloting, were used on national samples of pupils at ages eleven, thirteen and fifteen.

They represent an interesting potential source of material for teachers, although it is important to note that some of the materials, particularly those developed for assessing category 6, employ extensive observation checklists for use in one-pupil-to-one-observer situations. Hence they are not usable as they stand for whole or even part classes.

The findings of the APU have led to some very important insights into levels of pupil competence in science process skills.[21] They have also revealed important information about the nature of the testing of science process skills some of which should be taken into account when devising our own assessments, e.g. for National Curriculum purposes. For example, we now know that pupils perform differently in a pencil and paper test of planning an investigation to the way they do when given the opportunity to handle the potential apparatus involved. Generally the latter task is found to be easier. We also know from APU work that performance involving reading scales of measuring instruments varies widely with slight variations in the task and whether a diagram, a photograph or an actual instrument is used.

Table 8.4 **APU framework for science assessment**

1.	Use of graphical and symbolic representation	Reading information from graphs, tables and charts Representing information as graphs, tables and charts	Written test
2.	Use of apparatus and measuring instruments	Using measuring instruments Estimating physical quantities Following instructions for practical work	Group practical test
3.	Observation	Making and interpreting observations	Group practical test
4.	Interpretation and application	I Interpreting presented information II Applying: Biology concepts Physics concepts Chemistry concepts	Written test
5.	Planning of investigations	Planning parts of investigations Planning entire investigations	Written test
6.	Performance of investigations	Performing entire investigations	Individual practical test

National Curriculum assessment

At present teachers make National Curriculum assessments on their pupils at KS3 that are formally reported. From what can be gathered from the recently published Final Dearing Report this will continue in the future. Currently teacher assessments are made in all four attainment targets. At KS4 teacher assessment of Sc1 occurs that is incorporated into GCSE results. Voluminous guidance about making National Curriculum assessments has issued from SEAC (now SCAA) over the past four years. There is no indication that this will cease in the near future!

Teachers have found relatively few problems in assessing National Curriculum levels of attainment in Sc2 (Life and living processes). It has proved to be reasonably straightforward to align performance on conventional pencil and paper tests with National Curriculum levels. The Standard National Tests have illustrated how this is done, and indeed questions from these tests have quickly found their way into use as schools tests. Assessing Sc1 has been much more of a problem, particularly at KS4 when the highest levels, 7 to 10, are involved.

Before looking at the assessment of Sc1 in some detail, it is appropriate to stress two key points. First, assessment cannot be conducted on what is not there – the scheme of work must provide opportunity for the learning outcomes, represented by the statements of attainment of Sc1 to be developed by presenting appropriate teaching situations, as well as providing situations in which the assessment can be made. It would be entirely inappropriate if pupils were to be assessed on Sc1, just once, near the end of a key stage, in order to satisfy reporting requirements. Investigations for Sc1 arise from normal classroom activities (some examples were given in Chapter 3); they are not mini-practical examinations. Hence it is more fruitful to look for investigation possibilities in your own scheme of work rather than in trying to graft them in from other people's collections. The second point to make is about classroom organization and resource distribution. For at least part of the investigation pupils will be working in their normal groups. The important principle is that working arrangements should facilitate the autonomous activity of the pupils as far as possible. This frees the teacher from having to deal with minor organizational and administrative tasks and allows time to be spent on talking with, listening to and watching pupils. The following check-list may be useful:

1. The layout of the laboratory and its furniture should encourage cooperative work as far as possible – there should be enough space for equipment, books and for sitting together for discussion.
2. Everyday equipment (e.g. glassware) and resources (paper, felt-pens, glue) should be easily accessible to the pupils by being stored and labelled appropriately.

3. Pupils should have acquired good working habits:
   (a) knowing that they are expected to wash and replace glassware, etc.;
   (b) leaving working surfaces clean, dry and tidy;
   (c) moving around the lab carefully and purposefully;
   (d) behaving safely at all times (e.g. using goggles).
4. Teachers should ensure that pupils have clear targets about their investigation, for example: knowing what they are investigating, how long they have to complete the work, how the work is to be presented and reported.

Assessing Sc1

Attainment target 1 consists of three closely related strands:

Strand i: ask questions, predict and hypothesize
Strand ii: observe, measure and manipulate variables
Strand iii: interpret results and evaluate scientific evidence

SEAC have summarized the interrelationship between strands as follows:

> The nature of the question, prediction or hypothesis in strand (i) will determine what is observed or measured in strand (ii). These outcomes will enable pupils to interpret and evaluate their observations in strand (iii).[22]

What follows are guidelines for a strategy for making Sc1 assessments. They are based on the questions most often raised by new teachers when confronted by Sc1 and a synthesis of advice from a variety of sources.[23]

Guidelines for assessing Sc1

1. Any assessment undertaken of Sc1 must be within the context of a whole investigation, i.e. one in which all three strands are addressed. All three strands, however, need not be assessed within the one investigation. Pupils will need to be taught the component skills of Sc1. Such teaching does not have to be within a whole investigation although clearly sometimes that should be involved. We may decide that a particular piece of work is very useful for interpreting results and evaluating evidence (strand iii), or that another is a good exercise for developing observation and measurement skills (strand ii). This is fine but assessment of these skills, for Sc1 purposes, cannot be made in these separate contexts. One other point about pupils' performances on separate tasks of

this sort is that they are likely to be different from those displayed in the context of whole investigations. In a review of APU findings, Black[24] points out that many pupils who were able to use measuring instruments and procedures in isolated contexts failed to do so when required in a complete investigation. Asking a pupil to make a measurement from a given piece of apparatus is a different task from recognizing an appropriate setting for undertaking such a measurement. This involves a broader understanding of when, what and how to measure. Assessing process skills in the context of whole investigations (a 'holistic' approach) as opposed to doing so in separate activities ('atomistic') has been the subject of extensive debate. The strategy required for assessing Sc1 is that the investigation must be holistic but the assessment may be atomistic, focusing on one or more strands.[25]

2. Investigations will arise from work conducted in Sc2, 3 and 4. This is important because the level of sophistication of the investigation depends on the scientific knowledge and understanding or theory that has been previously developed and that is brought to the investigation, e.g. in Sc1 strand i:

   3a: suggest questions, ideas and predictions *based on everyday experience*, which can be tested.

   4a: ask questions, suggest ideas and make predictions, *based on some relevant prior knowledge*, in a form which can be investigated.

   5a: formulate hypotheses where the causal link is *based on scientific knowledge, understanding or theory*.

   For example, in an investigation of the role of yeast in bread-making coming as part of a unit of work on microbes, pupils might speculate that yeast must be present to make the dough rise (level 3a). With a fuller Sc2 background, a pupil might suggest that since yeast is a living organism it is its production of carbon dioxide that causes the dough to rise (level 4a). Another pupil might go further to suggest that since glucose is involved in respiration, adding glucose to dough might cause a greater effect (level 5a). A wide range of similar examples is provided in the publications referred to earlier.

   The level of performance in an investigation will, almost certainly, vary from context to context because of different levels of understanding of concepts and motivation. Because of this assessments will need to be made in a wide range of different content areas and hence they should be fairly frequent activities.

3. It is important that the investigations that are presented to pupils allow them to demonstrate achievement – hence they should be

INVESTIGATION SHEET

Your investigation

On your own

1. Think of a question to test.
2. Write an outline plan for your investigtion. Use the planning sheet to help you.
3. Say what you think you will find out and why.

In your group

Decide which plan to follow. Check with your teacher whether you can start your investigation.

Get all the equipment you need and carry out your investigation.
You may change your plan as you go along, but note any changes you make.

On your own

Use the report sheet to help you write up your investigation.

PLANNING SHEET

Your plan

In planning your investigation, think about these things:

- what you are trying to find out
- what you think will happen
- why you think this will happen using scientific ideas
- what you will measure/observe
- what you will alter between measurements/observations
- what you will keep the same to make it a fair test
- how many measurements/observations you will take
- how you will space out your measurements/observations
- what instruments you will use to make your measurements/observations
- safety problems that could arise

REPORT SHEET

Your report

In your report, think about including:

- what you were trying to find out
- what you altered
- what you measured or observed
- how you made your investigation a fair test
- what instruments you used to make your measurements
- your results, eg in a table
- a graph if suitable
- a description of any pattern
- a conclusion which matched your results
- a comment on how you could have improved your investigation
- an explanation of your results using scientific ideas

Figure 8.7 Another example of a planning sheet

Figure 8.8 Example of pupil-friendly Sc1 statements

accessible for them but they should not limit potential attainment. SEAC[26] has produced a useful check-list for screening potential investigations based on the openness of the investigation and the readiness of pupils to tackle different levels of work in the three strands.

4. One source of evidence of attainment will be the written reports that pupils make of their investigations. It is useful to provide them with a structure for such work in the form of planning or prompt sheets. One example of a planning sheet was given in Chapter 3; another is shown in Fig. 8.7. Teachers have given a lot of thought to the problem of communicating the key aspects of National Curriculum assessment to pupils so that they can become active participants in the process. One approach is to translate Sc1 statements into more pupil-friendly language. One example is shown in Fig. 8.8. Another interesting development is the use of self-assessment sheets by pupils, which can then form the basis for the records of achievement. The example shown in Fig. 8.9 comes from SEAC.[27]
5. In addition to written evidence, valuable information about pupils' progress can be gained on the spot as they conduct their investigations. This will come from discussions held between the teacher and an individual pupil or a small group. Such evidence of attainment has come to be called 'ephemeral'. Gathering such information is an important part of the Sc1 assessment strategy because pupils may know and be able to do far more than they are able or prepared to write down. Answers to questions such as: What are you going to do? Why are you going to do that? What do you think will happen? Why do you think that? etc. may result in you re-evaluating the opinion of attainment that you may have arrived at from written work. Such information, with a note of date and context, can be filed along with samples of written work to provide a portfolio of work that may be required for end-of-key stage audit (moderation) purposes.
6. There is no requirement that all pupils in the class are assessed in each investigation. Hence in one activity a particular group of pupils might be targeted by the teacher. This raises one frequently discussed issue. If pupils are working together as a group, how can the teacher be sure whose work is being assessed? Part of current received wisdom on this is that one's experience of the level at which each pupil is working accumulates through these extensive encounters, and written work and oral evidence build a picture as the pieces do of a jig-saw puzzle. Rather more pragmatically, many teachers ask pupils to work individually on generating ideas, raising questions, making predictions, etc, then to work collaboratively to conduct an agreed-upon investigation, and then individually again to draw conclusions and to record them. A

detailed account of a school's experience in working in this way is given in the OPENS project.[28]

Key stage 4

Teachers are responsible for assessing Sc1 for external GCSE purposes. This has a weighting of 25 per cent in all GCSE science examinations. The means by which this will be done have been covered in general terms in the previous account. There are minor variations in the approaches taken by the various GCSE boards but all are required to conform to the *GCSE/Key Stage 4 examination criteria* (SEAC, 1992). Taking the Southern Examining Group (Double Award) Syllabus as an example, we find Sc1 presented as shown in Fig. 8.10. The three strands have been named (planning, etc.) and the ten levels of each have been given a mark weighting of 1. Assessment of the three strands results in aggregation of marks to produce an overall score out of 30. Levels are assumed to be hierarchical, so a pupil reaching level 6 in all three strands would be awarded 18 marks. Such marks are subject to moderation by the board. Very helpful syllabus support material[29] includes examples of pupil investigations and how they have been marked.

For the foreseeable future, GCSE results are to be reported as an eight-letter scale (A* to G).

Statutory reporting requirements

Reporting pupils' achievements to parents is now a legal requirement of teachers. Schools must send parents at least one written report every school year. In such reports there is a required minimum of information about:

1. the pupil's progress in all National Curriculum subjects studied;
2. progress in all other subjects and activities;
3. general progress and an attendance record;
4. where relevant, the pupil's National Curriculum assessment results and how these compare to results of pupils at the same age in school;
5. any public examination results; and
6. the arrangements to discuss the report with the school.

Further details on all these areas are available in *DfE Circular 16/93: Reports on Pupils' Achievement* (December 1993). This document defines, in effect, a school code of practice. It will not be your responsibility to implement this but to work within it in ways that will be made clear by your Head of Department or curriculum leader. Your personal responsibility is to ensure that the system of assessment and

Name

Date:

Form

Investigation:

In this investigation I have shown that I can:

	Tick	Comments
• think of a question to investigate		
• predict what might happen in an investigation		
• give a reason for my prediction		
• decide what to change		
• decide what to measure		
• choose the best measuring instruments		
• make measurements		
• repeat measurements tor accuracy		
• describe what a fair test is		
• carry out a fair test		
• do my investigation safely		
• record my results clearly in a table		
• draw and label a bar graph		
• draw and label a line graph		
• understand when to choose a line graph		
• get information from my graph		
• describe patterns in my results		
• say what my results mean		
• recognise that my results are only useful if I have done a fair test		
• decide whether other people should believe my results		
• explain my conclusions using scientific ideas		

Figure 8.9 Pupil self-assessment sheet

record keeping that you adopt is appropriate to provide the information required, in the form needed, to produce such reports.

Assessment in the post-sixteen curriculum

Assessment of practical work by teachers at A-level has become widespread since the first introduction of a scheme in 1976 by the then JMB. Since that time virtually all the A-level boards have followed suit although a few have retained a practical examination, in some cases as an alternative to teacher assessment. Hence many teachers have extensive experience of this process. Most of the schemes for teacher assessment of practical skills are very similar. They involve making judgements about pupils' performance on a small range of skills (perhaps three or four) by means of a numerical scale (e.g. 0–6 or 1–10), each point of which is defined (perhaps) by a number of descriptors. Some of the issues that can arise have been discussed in the earlier section of this chapter on CR. The examination boards give detailed advice about the numbers of such assessments that should be made, whether assessments should be averaged or the highest attainments selected, and how marks are to be processed and recorded for submission to the board concerned.

In England, Wales and Northern Ireland teaching began in September 1994 on new A and AS syllabuses, which will be examined for the first time in 1996. The new syllabuses have a common subject core laid down by SEAC (July 1993), comprising 35 per cent and 70 per cent of A and AS syllabuses respectively. The core is designed to build from the knowledge and understanding described by level 7 in Sc2 of the 1991 Science Order. It specifies the skills needed to carry out experimental work in biology, encompassing investigative and illustrative practical work.

These are as follows:

Planning

Applying knowledge and understanding of biological principles, theories and models to develop hypotheses and make predictions.
Planning investigations including: identifying and ordering the steps involved; identifying key variables; deciding which variables to control; selecting appropriate apparatus and materials.

Implementing

Using apparatus and materials in an appropriate way.
Carrying out experimental work in a methodical and organized way with due regard for safety and for living organisms.
Making detailed observations; taking measurements to an appropriate degree of precision; making suitable drawings.

	Strand A Predicting	Strand B Implementing	Strand C Concluding
Level 1	(1 mark)	observe familiar materials and events (1 mark)	(1 mark)
Level 2	ask questions such as "how...?", "why...?", and "what will happen if...?", suggest ideas and make predictions. (1 mark)	make a series of related observations. (1 mark)	use their observations to support conclusions and compare what they have observed with what they expected. (1 mark)
Level 3	suggest questions, ideas and make predictions, based on everyday experience, which can be tested. (1 mark)	observe closely and quantify by measuring using appropriate instruments. (1 mark)	recognise their conclusions may not be valid unless a fair test has been carried out. distinguish between a description of what they observed and a simple explanation of how and why it happened. (1 mark)
Level 4	ask questions, suggest ideas and make predictions, based on some relevant prior knowledge, in a form which can be investigated. (1 mark)	carry out a fair test in which they select and use appropriate instruments to measure quantities such as volume and temperature. (1 mark)	draw conclusions which link patterns in observations or results to the original question, prediction or idea. (1 mark)
Level 5	formulate hypotheses where the causal link is based on scientific knowledge, understanding or theory. (1 mark)	choose the range of each of the variables involved to produce meaningful results. (1 mark)	evaluate the validity of their conclusions by considering different interpretations of their experimental evidence. (1 mark)
Level 6	use scientific knowledge, understanding or theory to predict relationships between continuous variables. (1 mark)	consider the range of factors involved, identify the key variables and those to be controlled and/or taken account of, and make qualitative or quantitative observations involving fine discrimination. (1 mark)	use their results to draw conclusions, explain the relationship between variables and refer to a model to explain the results. (1 mark)
Level 7	use scientific knowledge, understanding or theory to predict the relative effect of a number of variables. (1 mark)	manipulate or take account of the relative effect of two or more independent variables. (1 mark)	use observations or results to draw conclusions which state the relative effects of the independent variables and explain the limitations of the evidence obtained. (1 mark)

Figure 8.10 Grid to show the relationship between strands and statements of attainment

	Strand A Predicting	Strand B Implementing	Strand C Concluding
Level 8	use scientific knowledge, understanding or theory to generate quantitative predictions and a strategy for the investigation. (1 mark)	select and use measuring instruments which provide the degree of accuracy commensurate with the outcome they have predicted. (1 mark)	justify each aspect of the investigation in terms of the contribution to the overall conclusion. (1 mark)
Level 9	use scientific theory to make quantitative predictions and organise the collection of valid and reliable data. (1 mark)	systematically use a range of investigatory techniques to judge the relative effect of the factors involved. (1 mark)	analyse and interpret the data obtained, in terms of complex functions where appropriate, in a way which demonstrates an appreciation of the uncertainty of evidence and the tentative nature of conclusions. (1 mark)
Level 10	use scientific knowledge, and an understanding of laws, theories and models to develop hypotheses which seek to explain the behaviour of objects and events they have studied. (1 mark)	collect data which are sufficiently valid and reliable to enable them to make a critical evaluation of the law, theory or model. (1 mark)	use and analyse the data obtained to evaluate the law, theory or model in terms of the extent to which it can explain the observed behaviour. (1 mark)
	(10 mark)	(10 mark)	(10 mark)
MAXIMUM OVERALL TOTAL = 30 MARKS			

Figure 8.10 (Continued)

Concluding

Assessing the reliability and accuracy of experimental data.
Recognizing and commenting on trends and patterns in data.
Applying simple statistical tests (including the assigning of confidence levels where appropriate) to experimental results.
Drawing appropriate conclusions from experimental results.
Evaluating the validity of experimental activities and their conclusions.
Communicating biological information and ideas in appropriate ways, including tabulation, line graphs, histograms, continuous prose, annotated drawings and diagrams.

It will not have escaped your notice that the three groups above correspond to the three strands of Sc1 in the Science Order.

Assessment objectives or learning outcomes are derived from the skills and are described thus:

Students should be able to:

a) devise and plan experimental activities;
b) select and use appropriate techniques to carry out experimental activities;
c) interpret, evaluate and communicate the result of their experimental activities.

Currently, draft syllabuses for 1994 are available. The approach taken to assessing these experimental skills by one board is shown in Table 8.5.

The learning outcomes have been grouped under four headings for each of which six assessment criteria have been identified, to be marked on a scale of 0–2. Teacher assessment is based on routine course-work exercises (for three skills) and on an extended investigation. This is expected to be completed in twenty to twenty-five hours of class time, about fifteen hours of private study and takes the form of a report of 2500 words maximum length. It forms the basis for assessment of four skills (three the same as in the other teacher-assessed component).

What examination boards do *not* decide for you (either at present or in the future) is how you come to make the required judgements of your students. For example, in the scheme described above when is a criterion 'completely met' as opposed to being 'partly met'?

Two different approaches can be employed to help you make criterion-related decisions:

1. *checklists*, by which working procedures can be assessed on task;
2. *mark schemes*, which can be applied to written responses on completion of the task.

Mark schemes will be more useful in assessing planning and concluding sub-skills, check-lists for implementing ones. Two examples will illustrate (both in this case from 'implementing').

Example (a)

Learning outcome: 'Apparatus and materials are used in a careful and well-organised way'.

Context: The use of a simple potometer to investigate the effect of environmental variables on the rate of water uptake of a cut shoot.

Check-list: 1. Cutting the stem and fitting it to the capillary tube under water.
2. Blotting off any water from the leaves.

Table 8.5 **Assessment of experimental skills**

		Assessment by teachers	
Skills	*Assessment criteria*	*Practical exercise*	*Extended investigation*
A. Planning	Application; definition; variables; organization; apparatus; procedures	*	*
B. Implementing	Carrying out; safety; manipulative skills; observations (accuracy and detail); observations; (suitable format); precision/accuracy	*	*
C. Interpreting and concluding	Analysis; reliability modification; interpretation; conclusion; communication	*	*
D. Researching	Preliminary research; variety of sources; selection; communicating information; referencing bibliography	–	*
	Raw total	36	48
	Exam total	10%	10%

* context in which the skill is assessed
– not assessed in this context

3. Clamping at an appropriate point on the stem.
4. Operating the variables appropriately e.g. Bunsens do not singe leaves, polythene hoods do not touch leaves; droplet sprays do not soak leaves.
5. Manipulating the bubble sensibly in the capillary tube – to make readings over a section of the scale near the bottom of the capillary tube; ensuring the bubble does not escape into the rubber connector.

You will have observed the students during the activity and have noted their performances relative to the check-list. You still have to make a judgement, which inevitably is subjective, about their mastery levels. Questions that will come to mind are:

What level of expertise can I reasonably *expect* from these students?
Within this time allocation?
Doing this for the first time?
Should they perform all of the operations correctly without obvious transgression?

My judgements would be as follows:

For 2 (criterion completely met): all of the above
For 1 (criterion partly met): three of the five points
For 0 (criterion not met at all): fewer than three.

Yours might be different; the moderation process will arbitrate!

Example (b)

Learning outcome: 'Make accurate and detailed observations including low and high power drawings of a specimen.'

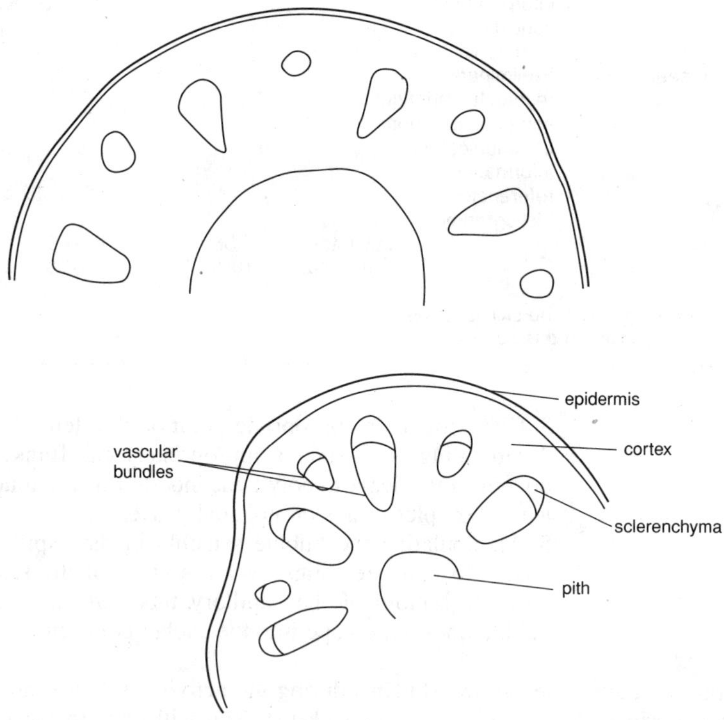

Figure 8.11 Drawings of a transverse section of a buttercup stem by two A-level students

Context:	Make a labelled plan drawing of half of the stained transverse section of part of a flowering plant. Accurately show its shape and the distribution of the different tissues in the section.

A detailed marking scheme is applied as follows (the specimen is a transverse section of a buttercup stem):

Drawing marks	*Labels*
Ratio of cortex : pith is less than 1;	epidermis;
presence of a thin epidermis;	vascular bundle;
occurrence of two sizes of vascular bundles;	cortex;
large and small bundles alternate;	pith.
their outer margins occur on the same arc;	
smaller bundles are less than half the size of larger; between five and eight large bundles shown in the half-section.	
Total: 7 marks	Total: 4 marks

Figure 8.11 shows the responses of two A-level students. What marks would you award to each? Applying the marking scheme to these responses results in a mark, in this case, out of 11. Again you have to decide on the appropriate mark score relative to the performance criteria.

Exactly the same procedure can be used to assess physiology or biochemical investigations where observations of events are involved, e.g. changes in colour, size, shape, mass, bubbling rate, etc. and the interpretations or conclusions made from the observations.[30]

Moderation

Using procedures of the kinds discussed above, it is your responsibility to produce a reliable rank order of the attainment of your students. What you cannot be expected to know is how your judgements would compare with those of another teacher. You might be strict or lenient. Particularly if you work in a school with a small sixth form, in any one year you may not see work that represents a wide range of grades. Examination boards recognize this and all teacher assessments undergo a process of moderation.

Statistical moderation

This involves making appropriate adjustments to teacher-assessed scores by reference to the performance of candidates on a moderating instrument, i.e. a paper or a group of papers that is taken by all candidates. The thinking behind this is that if the mean mark of your candidates on the moderating instrument is similar to that of the whole entry, then the mean mark of your students' practical scores would be expected to be close to the mean practical mark of the entry. If this is not the case then the practical marks are adjusted by an amount necessary to bring them into line. This procedure does not change the rank order of your candidates on their practical scores.

Moderation by inspection

As the name suggests, this involves sending samples of students' work to the examination board for remarking to occur. Following such remarking, the teacher's marks would be confirmed or scaled by an appropriate factor. Again the rank order of candidates remains unchanged.

Both methods of moderation have technical, logistical and financial implications (not to say flaws) but these are largely irrelevant for you because you will work within the system operated by your chosen board. Nevertheless it is very useful to know how your marks are moderated, and the extent and reasons for any adjustments made to them so that you can begin to get a feel for expected standards of performance of students.

Marking pupils' on-going class work or homework

Most of the material in this chapter so far has involved dealing with issues that arise in the context of conducting formal assessments where devices such as tests, etc. are used for the sole purpose of making assessments. These are, or at least should be, intimately bound up with the process of teaching. As well as this, and even more central to teaching, are the various written tasks that we ask pupils to perform, often for a variety of purposes. These were reviewed in an earlier chapter. Traditionally how we deal with work of this sort is regarded as marking rather than assessing. Despite this, marking pupils' work does have an implied assessment dimension and is certainly perceived as such by pupils. It is such an important dimension of the teacher's role that it is worth giving a little attention to some of the issues raised in doing it.

Philip Benton[31] lists a range of possible options by which teachers might respond to pupils' written work:

1. going through the work correcting all errors of grammar, spelling and punctuation;
2. selecting particular errors for correction;
3. correcting misunderstandings or factual errors;

4. giving a mark or grade;
5. adding a written comment;
6. making a note of particular problems to raise subsequently with the pupil;
7. suggesting/requiring corrections to be done, including ones of spelling;
8. going over common areas of difficulty with the whole class;
9. reading out particular pieces of work;
10. seeing individuals about their work;
11. pinning work up as a display;
12. simply putting a tick or date to indicate that it has been read;
13. using it as the basis for further work.

The options in this list are not mutually exclusive; as a teacher you would find yourself using perhaps three or four of them for most of your marking. One or two of them are so deeply ingrained into our collective notions of how teachers operate that they hardly seem to require further thought or comment. Nevertheless there are certain general principles that do require further comment.

This is another of those areas where you will be able to take advantage of the existence of a departmental or school marking policy; written guidelines should be available about the mechanics of correcting work. If not, your colleagues in the English department will be able to offer you advice.

Clearly, the way in which you decide to mark a piece of work is related closely to the purpose for which it was set and the nature of the task itself. Earlier we distinguished between transactional and expressive writing. A report of an experiment or a summary of a topic are, essentially, transactional. Here the emphasis is on factual accuracy and completeness of the ideas concerned, perhaps with a fair sprinkling of scientific terms. If you find inaccuracies or omissions, it is pointless to do anything other than draw these to the attention of the pupil. Most people would do this by writing in the margin, in the text itself or at the end of the piece of work. That's fine, but beware of overdoing it. Pupils complain bitterly about work handed back by the teacher that is covered in red ink. As to the pros and cons of marks and grades (still hotly debated among teachers) what *purpose* is served by such devices? From your point of view A may be excellent, C mediocre, etc., but what does this mean to the pupil? Similarly there is only point in giving a mark of 7 out of 10 on the page of the pupil's exercise book if he or she knows what else they would have to have done to have gained 10 out of 10. So, how are you going to arrange to provide that information? Some teachers inform their classes of the meanings they attribute to marks:

> *Your work will be marked out of 10*
> A mark of 8–10 means that you have included most or all of the

key points required.
5–7 means that some important points have been missed.
Below 5 shows that most of the important points are missing and the work needs considerable attention.

This allows teachers to list the missing ideas at the end of the piece of work, perhaps a slightly idealistic recommendation in the light of the mountain of exercise books to be confronted!

Spelling errors should be corrected; apart from anything else, in all GCSE examinations 5 per cent of the marks available for written papers are now allocated to spelling, punctuation and grammar. It does not matter too much how correction is done except that indicating that the word is misspelt is not enough – either the correct version must be provided or the pupil must find it out by consulting a dictionary. Most teachers take the line of least resistance and write the word out for the pupil! Be cautious too about your efforts to help, which may turn out to be counter-productive. Asking the pupil to write out a particular term five times might be intended to benefit the pupil; the pupil however construes it as punishment and resents it.

Expressive writing emphasizes the value of the personal contribution and obviously demands a response in kind. A poem, a letter written to a friend, etc. requires very little correction, if any. Some suggestions for redrafting or an expression of pleasure given to the reader are more appropriate responses, followed by sharing the writing with a wider audience, e.g. by displaying it.

It is also possible to relate written work to National Curriculum attainment target levels and base feedback on these. This is appropriate when this nomenclature has some meaning for the pupils. It might take considerable time and effort to develop these shared meanings. As one pupil put it when given some feedback about his ability to control variables in an investigation, 'Yes, but I'd rather have a mark out of ten, I can understand that.'

Finally, some cautionary thoughts indicating the need to guard against bias in your responses to pupils' written work. The presentational level of pupils' work varies enormously, particularly their handwriting. We want to encourage pupils to present work in a way that is easy to understand and looks attractive. We may commend them in one way or the other for doing so or for making a particular effort to do so. But do not base your judgement of the value of the content of such work on what it looks like. And, yes, there is research evidence[32] that shows that when presented with the same piece of written work copied out into *different* styles of handwriting, experienced teachers were found to be biased towards particular styles, awarding higher marks to some.

There is also evidence of gender bias in teachers' responses to pupils' work in science. In a study by Goddard-Spear[33] teachers marked written work that was attributed to a boy or a girl and then the work was remarked with the attributions reversed. Work attributed to a boy was rated more highly than when it was presented as having been performed by a girl over a range of different criteria. Such bias was most marked for female teachers marking boys' work (most lenient) and when male teachers marked girls' work (most severe). This is another of these 'forewarned, forearmed' situations for you.

Notes and references

1. SSCR (1987) *Better Science: Assessing progress, Curriculum Guide 11*, p. 18. London: Heinemann Educational Books.
2. Further details are available in standard texts on educational testing, e.g. Ebel, R. L. (1972) *Essentials of Educational Measurement*. New Jersey: Prentice-Hall.
3. Ebel, R. L. (1972) ibid.
4. Lindquist, E. F. (1951) *Educational Measurement*. Washington, DC: American Council on Education.
5. Brown, S. (1980) *What Do They Know*? Edinburgh: HMSO.
6. 'Can do' styles of reporting have been involved in a number of recent developments including *Suffolk Science*, OCEA and GASP.
7. From TAPS 1 (Techniques for the Assessment of Practical Skills) published by Heinemann Educational Books. The complete resource includes TAPS 2 and 3.
8. Wood, R (1991) *Assessment and Testing. A Survey of Research*, p. 102. Cambridge: Cambridge University Press.
9. Simpson, M. (1985) Diagnostic assessment and its contribution to pupils' learning, in Brown, S. and Munn, P. (eds) *The Changing Face of Education 14–16: Curriculum and Assessment*, pp. 69–81. London: NFER-Nelson.
10. Haslam, F. and Treagust, D. F. (1987) Diagnosing secondary students' misconceptions of photosynthesis and respiration in plants using a two-tier multiple choice instrument. *Journal of Biological Education*, vol. 21, no. 3, pp 203–11.
11. Most recently in Sutton, C. (1992) *Words, Science and Learning*, chapter 8. Buckingham: Open University Press.
12. Biological Sciences Curriculum Study (1990) *Biological Science: A Molecular Approach*. Lexington, MA: DC Heath.
13. The most accessible starting point for information on concept mapping is Novak, J. D. and Gowin, D. B. (1984) *Learning How to Learn*. Cambridge: Cambridge University Press.
14. This adaptation of Novak's method is in Stuart, H. A. (1985) Should concept maps be scored numerically? *European Journal of Science Education*, vol. 7, no. 1, pp 73–81.

15. Readable texts are: Frith, D. S. and Macintosh, H. G. (1989) *A Teacher's Guide to Assessment*. Cheltenham: Stanley Thornes; Satterley, D. (1989) *Assessment in Schools*, 2nd edn. Oxford: Blackwell; and Macintosh, H. G. (ed.) (1974) *Techniques and Problems of Assessment*. London: Edward Arnold. For a text which concentrates on issues in science assessment see Kempa, R. (1986) *Assessment in Science*. Cambridge: Cambridge University Press.
16. Murphy, P. (1991) Assessment and gender. *Cambridge Journal of Education*, vol. 21, no. 2, pp 203–14.
17. See Vance, M. (1984) *Assessment in a Multicultural Society: Biology at 16+*. Longman for the Schools Council.
18. SSCR (1984) *Towards the Specification of Minimum Entitlement: Brenda and Friends*. London: Schools Council.
19. SSCR (1987) op. cit.
20. Fairbrother, R. (1989) Problems in the assessment of scientific skills, in Wellington, J. (ed.) *Skills and Processes in Science Education: A Critical Analysis*, chapter 6. London: Routledge.
21. See Russell, T. P. *et al.* (1988) *Science at Age 11: A review of APU findings 1980–84*; Schofield, B. *et al.* (1988) *Science at Age 13*, etc.; and Archenhold, W. F. *et al.* (1988) *Science at Age 15*, etc. All published by HMSO.
22. SEAC (1993) *School Assessment Folder Assessing Sc1 Scientific Investigation*. p. 5.
23. Apart from SEAC (1993) ibid, see *1992 KS3 School Assessment Folder (Part 3) March 1992 Materials to support the assessment of Sc1*; SEAC (1993) *Pupils Work Assessed (Science) Four pupils' folders*; Jones, A. T. *et al.* (1992) *Open Work in Science Development of Investigations in Schools*. Hatfield: Association for Science Education. Advice on Sc1 assessment is also available from the ASE and individual LEAs.
24. Black, P. (1990) APU Science – the past and the future. *School Science Review*, vol. 72, no. 258, 13–28.
25. Jones, A. T. *et al.* (1992) OPENS project, op. cit.
26. SEAC (1993) op. cit., pp. 24–5.
27. SEAC (1993) ibid., p. 27.
28. Jones, A. T. *et al.* (1992) op. cit.
29. *National Curriculum Science Syllabus Support Material* (1992) from Southern Examining Group, Publications Department, Stag Hill House, Guildford, Surrey, GU2 5XJ.
30. Although predating the 1994 syllabus changes, a useful source of ideas is Sands, M.K. and Bishop, P. E. (1984) *Practical Biology: A Guide to Teacher Assessment*. London: Bell and Hyman.
31. Benton, P. (1981) Writing: how it is received, in Sutton, C. *Communicating in the Classroom*, Chapter 1. London: Hodder and Stoughton.
32. Briggs, D. (1980) A study of the influence of handwriting upon grades using examination scripts. *Educational Review*, vol. 32, no. 2, 185–92.

33. Goddard-Spear, M. (1984) The biasing influence of pupil sex in a science marking exercise. *Research in Science and Technological Education*, vol. 2, no. 1, 55–60.

CHAPTER 9

Biology across the curriculum

The whole taught curriculum

We could consider this to be the totality of the learning experiences offered by the school. This is a useful concept because since we operate mainly as subject teachers in secondary schools we are largely concerned with the organization and teaching of that subject. As biologists we are also much occupied with connecting ideas up with those in chemistry and physics. What we might overlook is that the situation is quite different for our pupils. They leave school at the end of a day or indeed at the end of their school careers having been exposed to the combined battery of these subject specialists! It is important that all this adds up to a meaningful set of experiences for them – an idea often summed up by the notion of 'coherence'. The separate bits of the curriculum should add up to make a logical whole.

Section 1 of the 1988 Education Reform Act required schools to provide a broad and balanced curriculum. The National Curriculum that was enacted was established on the basis of ten subjects. However, since each subject was planned by a different subject working group and emerged at a different time (an earlier version of science was the first to appear along with mathematics), it was difficult to see how the curriculum would ever take on the quality of coherence. Recognition of this difficulty came with the appearance of a series of NCC Curriculum Guidance publications. One that set the context for a series of later publications and initiatives was *Curriculum Guidance 3 The Whole Curriculum* (NCC, 1990).

There were some over-optimistic thoughts in this document about the curriculum planning that might occur in school:

> In due course it is likely that schools will throw all the attainment targets in a heap on the floor and reassemble them in a way which provides for them the very basis of a whole curriculum. NCC, 1990, p. 1

That has not happened. Neither was it ever likely to do so because the numbers, content and status of the attainment targets have never been stable for long enough for them to function as bricks for curriculum construction in the way implied in that quotation. Some other areas of *Curriculum Guidance 3* have not stood the test of (comparatively short) time either. The idea that the National Curriculum would be a common curriculum compulsory for all pupils has been overturned by the Final Dearing Report's intention to place a great deal of choice back into the curriculum.[1]

Other areas of *Curriculum Guidance 3*, such as the cross-curricular elements remain important. The functions of these were to 'tie together the broad education of the individual and augment what comes from the basic curriculum'. The cross-curricular elements involve three aspects: dimensions, skills, themes.

Dimensions are essentially about ensuring that the school is a decent place in which to teach and to learn, i.e. they involve climate or ethos. Pupils require equality of access to all areas of the curriculum. This should come about through the development of whole school policies on gender equality, recognizing cultural diversity and promoting special educational needs. These were laudable aims and ones to which many schools had made significant progress through their own initiatives and through those of HMI and LEAs.

Skills are transferable, independent of content and they could be developed in different contexts across the whole curriculum. Those specified in *Curriculum Guidance 3* were: communication, numeracy, study, problem-solving, personal and social, and information technology.

Five *themes* are identified as essential parts of the whole curriculum. They are: economic and industrial understanding, careers education and guidance, health education, education for citizenship,

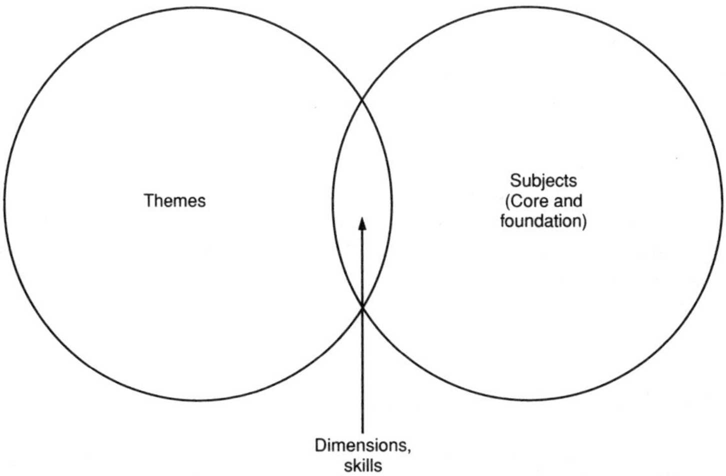

Figure 9.1 Relationships between the various components of the National Curriculum

and environmental education. Themes are rather more obviously content-based than dimensions and skills but, like them, were required (and still are) to appear somewhere in each school's curriculum.

The relationships between the various components of the National Curriculum are summarized in Fig. 9.1. Hence dimensions and skills are subsumed within both themes and subjects, addressed through the working methods employed in both. The degree of overlap of themes and subjects could be adjusted to vary considerably and as such forms the basis for a number of different models for curriculum planning. Five possible models were reviewed in *Curriculum Guidance 3* based on whether the themes would be taught:

1. entirely separately from subjects;
2. entirely through the National Curriculum subjects; or
3. by a combination of (1) and (2).

In most secondary schools approach (3) has evolved or is developing. This is due to the considerable content overlap between some of the themes and some National Curriculum subjects and to the existence of well-established programmes of personal and social education (PSE), which already incorporated elements of the themes. It is for this last reason also that I have chosen the structure of what follows in this chapter.

As a biologist you will find yourself dealing with aspects of the cross-curricular elements *as you teach National Curriculum Science* (in some instances this may not be obvious to you). It is perfectly possible to be teaching a topic that, whilst dealing with some aspect of the programme of study (PoS), also calls upon skills of communication, problem-solving or IT. If it is set in an industrial or environmental context, areas of one or more of the themes may be involved too. Your other potential role as a *biology* teacher is in supporting the cross-curricular aspects by making a teaching contribution in the PSE programme, perhaps as a form tutor. You will contribute in a number of other ways through your role as a teacher; however, in the rest of this chapter I will concentrate on your contribution as a biologist. For that reason I have restricted my attention to one dimension (multicultural perspectives), and to three themes (economic and industrial understanding, health education and environmental education).

Multicultural and anti-racist science teaching

One of the biggest challenges facing the science teacher is that of presenting science to pupils as a cultural activity that can be analysed and exemplified from a number of different perspectives. The school curriculum as a whole should demonstrate:

a commitment to providing equal opportunities for all pupils, and a recognition that preparation for life in a multicultural society is relevant to all pupils should permeate every aspect of the curriculum.

Also:

introducing, multicultural perspectives into the curriculum is a way of enriching the education of all our pupils. It gives pupils the opportunity to view the world from different standpoints, helping them to question prejudice and develop open-mindedness.[2]

As a core curriculum subject science should make a significant contribution to these aims.

The Non-statutory Guidance to the Science Order (NCC, 1989) makes clear what is expected of science teachers:

People from all cultures are involved in scientific enterprise. The curriculum should reflect the contributions from different cultures, for example, the origins and growth of chemistry from ancient Egypt, Greece and Arabia to the later Byzantine and European cultures, and parallel developments in China and India. It is important that science books and other learning material should include examples of people from ethnic minority groups working alongside others and achieving success in scientific work. Pupils should come to realise the international nature of science and the potential it has for helping to overcome racial prejudice.

This is supported strongly in the policy statement of the ASE *Race, Equality and Science Teaching* (1991), which includes among other recommendations for teaching and learning science that it should:

1. help pupils to question the reasons underlying inequalities between peoples and nations, and relate such issues to those of global interdependence;
2. question and challenge racist attitudes and assumptions;
3. provide positive images and avoid stereotypes of people, places and times;
4. show science as a worldwide human activity;
5. encourage a global awareness of environmental issues;
6. build on the cultural background and experiences of learners.

These aims involve two main interconnected sets of ideas that have become known as multicultural and anti-racist education. In a recent review, Dennick[3] points out that multicultural education in science might involve looking at the scientific contributions made by other cultures (past and present) and it should also stress the essentially similar processes of problem-solving that occur in different cultures. Anti-racist science education needs to go further – to challenge and oppose racism in society. Hence the areas of concern from this perspective are issues of race and IQ the geography of disease, exploitation of developing countries, etc.

Both approaches need to be put into operation by all science teachers. Nevertheless you will find science teachers who will argue that science is culture-free and it is this aspect of it which should be stressed to pupils. This may be put with most force by those working in schools with few pupils from minority groups. Chamberlain points out[4] the converse argument, held by the Swann Committee,[5] that since society is mobile, in the long term it is impossible not to be aware of and influenced by its multicultural nature. Hence failing to include these aspects could be seen as impoverishing the curriculum of all British pupils.

In the SSCR publication *Better Science: Working for a Multicultural Society*[6] some examples are provided of four approaches that groups of teachers have tried out in the science curriculum:

1. using science topics to consider the multicultural history of science;
2. making reference to contemporary practice in other societies;
3. drawing on the experience that some youngsters from ethnic minorities may have had personally or know about through their parents;
4. using topics, particularly from biology, to counter so-called 'scientific' racism.

A number of examples are given for each of these approaches, including biological ones.

The potential role of the biology teacher in work of this sort is enormous. One might reasonably claim that in almost every topic in biology the opportunity exists to embrace some aspect of this approach, to a greater or lesser extent.[7] And in one particular topic, that of human diversity, no one is better placed than the biology teacher for dealing with issues that arise. This does not involve tacking yet more content on to the scheme of work, but of undertaking a slight reorientation in it. For example, when dealing with food tests in a nutrition topic, you might include as many foods as possible, such as yam, sweet potato, ochra, peppers, lentils, rice, etc. Peacock[8] offers the following advice for avoiding implied racism in such a topic at primary level:

1. by not stereotyping some tastes/foods as better/worse;
2. by using community languages for naming/labelling breads;
3. by discussing why some foods are unacceptable, e.g. animal fats to vegetarians, etc.;
4. by avoiding books that only deal with 'our' kind of bread and baking;
5. by questioning culture-specific notions, e.g. 'tea-time'.

For some pupils such topics will extend their knowledge of other cultures; for pupils from ethnic minorities, they will represent areas of direct relevance for them that may result in greater motivation and in development of personal esteem. I have already indicated that as a biologist you are able to deal with aspects of human diversity in a way that your colleagues cannot. Michael Vance[9] maintains that in biology lessons providing information about human genetics and the interaction between inheritance and environment is vital to combat racism. He contends that even in biology, race is often treated as unproblematic and raises the question as to whether the term 'race' should be used at all. To me it seems necessary to use the term in order to show the inappropriateness of applying it to human beings. As Stephen J. Gould says:

> Human races are not separate species or ancient divisions . . . They are recent poorly differentiated subpopulations of our modern species, *Homo sapiens*, separated at most by tens or hundreds of thousands of years, and marked by remarkably small genetic differences.[10]

As a biology teacher you will want to deal with interesting examples of human variations such as skin colour, and the geographical variations in the distributions of the alleles for sickle-cell anaemia, thalassaemia and blood groups. Your role here is in helping pupils to appreciate that whilst differences between human populations exist, the diversity within such populations is very much greater. Gould quotes Richard Lewontin as follows:

> If, God forbid, the holocaust occurs and only the Xhosa people of the southern tip of Africa survived, the human species would retain 80 per cent of its genetic variation.

Your wider contribution to the school's PSE programme could well be in helping your colleagues to appreciate why the statement '*Homo sapiens* can be classified into a number of different races' is incorrect. They will know already of the historical examples where discrimination, persecution and genocide have followed from acceptance of the idea that such classification is possible. Back in your own classroom, you should be thinking about how the work you are going to do might be developed to question and challenge racist attributes and assumptions.[11] Similarly a good deal of thought is needed about the teaching methods to be employed. A considerable amount of advice exists about the teaching of controversial issues[12] in the curriculum. The role of the teacher varies with the nature of the issue that is to be explored. Three key concepts relating to the teacher's approach are: balance, neutrality and commitment. A balanced picture might be given when a range of alternative views about an issue are presented, e.g. the arguments for and against building nuclear power stations. The teacher's role may be to remain neutral throughout a discussion based on these arguments. Where a teacher indicates commitment to a particular idea, the stance of being neutral is abandoned in favour of one of persuasion. The danger of this is often said to be that of indoctrination. In any piece of work in which there is an attempt to incorporate anti-racist sentiments, clearly the only possible stance to be taken by the teacher is that of persuasion and we do not need to feel apologetic about that at all – we have the full weight of biological accuracy behind us!

Finally it is important that all your good work is not undone by the images conveyed to pupils by the written and other resource materials that we use in lessons. Ditchfield[13] provides the following useful check-list by which we may guard against this.

A check-list for analysing images in teaching materials and curriculum resources.

1. Examining images of science:
   (a) Is the application illustrated with examples from all over the world?
   (b) Is up-to-date technology only illustrated by Western examples?
   (c) Is science set in a context that relates to people?
   (d) Does the way in which the material is presented suggest that only Western ways are valid?
2. Examining images of people:
   (a) Are the people featured in the material from more than one cultural background?
   (b) Does the material present a positive view of people from different cultural backgrounds?
   (c) Are the negative images confined only to people from a particular cultural background?
   (d) Are people who are considered eminent scientists from different parts of the world represented?
3. General:
   (a) Is the material presented in such a way as to eliminate damaging feelings of superiority, based on race, in the European person?
   (b) Are the customs, life-styles and scientific traditions of people presented in a way that helps to explain the value, meaning and role of customs in their lives?
   (c) Are the images presented by the text valid?
   (d) What is the effect of these images on the attitudes of youngsters?
   (e) What are the ways that teachers compensate for deficiencies in the materials?

Economic and industrial understanding (EIU)

As a cross-curricular theme, EIU requires similar emphasis to the others listed in *Curriculum Guidance 3*. EIU is described as follows in that publication:

Education for economic and industrial understanding aims to help pupils make decisions such as how to organise their finances and how to spend their money. It involves controversial issues such as government economic policy and the impact of economic activity on the environment. It prepares pupils for their future economic roles as producers, consumers and citizens in a democracy. Pupils need to understand enterprise and wealth creation and develop entrepreneurial skills.

Education for economic and industrial understanding should cover aspects of industry and the economy such as business, commerce, finance and consumer affairs. Pupils should have direct experience of industry and the world of work and take part in small-scale business and community enterprise projects.

The knowledge, understanding, skills and attitudes that relate to EIU which pupils should develop are identified in two further NCC publications.[14] Some of the experiences that support these aims will be provided by pupils' work in science; others such as work placements, projects and mini-enterprises will be organized through whole-school programmes. You will find that teaching science through industrial applications and contexts can make it relevant and interesting for pupils. The NCC publication *Science and Economic and Industrial Understanding at Key Stages 3 and 4* (1992) illustrates how the Science Order provides a range of opportunities for developing EIU within the science curriculum. Some obvious starting points from the programmes of study for Sc2 (Life and living processes) are:

the uses of enzymes and microbes, for example, in the baking, brewing and dairy industries; how the body's natural defences may be enhanced through immunisation and medicines; beneficial products and services need to be balanced against human effects on the environment; They should consider the social, economic and ethical aspects of cloning and selective breeding. how food production involves the management of ecosystems . . . and . . . imposes a duty of care.

For biologists these areas represent familiar ground: they are all examples of biotechnology. This has been the focus of considerable attention in schools, and elsewhere, over the last ten years. A considerable literature exists in this field, which includes many suggestions for science teaching and which allows us to address EIU issues effectively. Some examples are discussed next.

Biotechnology in school science

Biotechnology is 'the application of biological organisms, systems and processes to manufacturing and service industries'.[15] The growth in this new technology in recent years has parallelled that of IT, and we find examples of it over a whole wide range of industries (Table 9.1).

Table 9.1 **Industrial biotechnological applications**[16]

Industrial sector	*Products/processes*
Agriculture	Animal feedstuffs Veterinary vaccines Manure and composting processes Microbial pesticides *Rhizobium* and other N-fixing bacterial inoculants Plant cell and tissue culture New plant varieties
Pharmaceuticals	Antibiotics Diagnostic agents (enzymes, antibodies) Steroids Vaccines
Food and drink	Dairy products (cheese, yoghurt, whey) Beverages (alcohol, tea and coffee) Baking Food additives Novel foods Glucose and high fructose syrups
Chemicals	Enzymes Polymers (mainly polysaccharides) Ethanol, acetone, butanol Organic acids (citric acid) Metal leaching
Energy	Ethanol (gasohol) Methane (biogas)
Environment	Water purification Effluent treatment Waste management Oil recovery

A range of curriculum changes and curriculum development initiatives (including GCSE, TVEI, TVEE and SSCR) have laid emphasis on the social, economic, technological and industrial dimensions of science and have drawn extensively on many of the areas listed in Table 9.1. These areas are equally relevant to science, technology and for addressing EIU. Probably the most obvious starting point, and the most accessible area for younger pupils, is biotechnology in the food industry. Table 9.2 shows the range of products involved that could be used for case study material.

Table 9.2 **Biotechnology and food**[17]

Traditional applications	*New applications*	*Food additives*
Cheese manufacture	Single cell protein	Citric acid
Yoghurt	Enzyme technology	Vanilla
Bread-making		Monosodium glutamate
Brewing		Xanthan gum
Vinegar production		Aspartame
Soy sauce		Cochineal

Many of these examples are useful, not only for addressing knowledge and understanding objectives but for open-ended, problem-solving activities. A simulation exercise, 'vinegar production', which was developed by the Sheffield Biotechnology Group,[18] illustrates a problem-solving approach very well. After an introductory activity pupils are required to design, cost, build and operate a fermentation tower. Their yield is assessed and the whole process is costed. An element of competition is introduced by identifying the winner as the company making the biggest profit. A wide range of teaching approaches can be built into these industrial case studies including discussion, simulation and role play. Another unit by the Sheffield SSCR group[19] involves yoghurt production. The unit involves four different activities and requires three to four hours teaching time. The structure of the unit is as follows:

Activity 1: How yoghurt is manufactured: a reading analysis including a DART exercise.

Activity 2: Making a natural yoghurt: a practical exercise involving production of natural yoghurt.

Activity 3: Making a nice yoghurt: assessing consumer preferences and identifying the ingredients in commercially successful yoghurt.

Activity 4: Designing a new yoghurt: a role play exercise that explores the decision-making necessary for the launch of a new commercial product.

Clearly work of this sort meets many of the EIU objectives mentioned earlier. A wide range of suggestions for other exercises of this type are available.[20]

Work experience

Curriculum Guidance 4 (NCC, 1990) states that pupils should also gain direct experience of the world of work: for younger pupils by occasional short visits; for older ones by work experience. Many schools have experience of working with local industry in Compact and other young initiative schemes. If you find that you are required to develop such contacts from scratch, obvious starting points for advice are parents, colleagues, LEA advisers and your local SATRO (Science and Technology Regional Organisation). Contacts with industry are also promoted through the Teacher Placement Scheme[21] by which teachers are able to spend varying lengths of time in a wide variety of business or industrial placements. For example, in a short placement of a few days in a local creamery, a teacher identified the following aspects of the functioning of the plant as providing case-study material that could be followed up by pupils:

1. studying the microbiological procedures for monitoring the quality of incoming raw milk;
2. identifying the procedures involved in the manufacture of soft cheeses;
3. determining what types of waste were produced, how they were treated and dispersed;
4. considering the interactions between the factory and the local community and environmental considerations, e.g. noise levels, traffic flow, smell and discharge into the local stream;
5. examining the management structure and identifying the roles and responsibilities of different personnel.

Health education

The World Health Organization defines health as a human right: not just the absence of disease but a state of total physical, psychological and social well-being. Health education provides guidelines for achieving that ambitious aim. Elements of health education have been part of biology teaching for many years. Recently such elements have been complemented, in school PSE programmes of active tutorial work, by topics such as 'drugs', 'relationships', 'smoking' and 'sex education'. Coming up to date, we find topics that are part of health education present in the National Curriculum Science Order. Health education is also a cross-curricular theme, and aspects of it have been legislated into the curriculum by the Education (No. 2) Act of 1986, the Education Reform Act of 1988, and the Education Act of 1993.

The key document by which the cross-curricular nature of health education is addressed is *Curriculum Guidance 5, Health Education* (NCC, 1990). That document concludes as follows:

Between the ages of 5 and 16, pupils should have received health education which:

- offers them opportunities to achieve their physical, psychological and social potential;
- helps them to know themselves better and to think well of themselves;
- provides appropriate health-related knowledge and helps pupils to develop the understanding and skills to use it;
- promotes attitudes and behaviour which contribute to personal, family and community health;
- promotes positive attitudes towards equal opportunities and life in a multicultural society by dealing with values and with cultural and religious beliefs.

As for the other cross-curricular themes, there are knowledge and understanding objectives involved as well as skills, attitudes and behaviours to be developed. A whole-school approach is essential to manage an effective programme in which the appropriate degree of overlap is attained, for example between subject teaching and work done in PSE; and to plan for the progression necessary to develop depth of coverage throughout the pupils' school careers.

Curriculum Guidance 5 identifies nine components of a health education curriculum from five to sixteen: substance use and misuse, sex education, family life education, safety, health-related exercise, food and nutrition, personal hygiene, environmental aspects of health education, and psychological aspects of health education.

In this document, these components are described more fully, and a framework is given that illustrates the appropriate areas of study for each key stage.

A number of the components listed above clearly have a biological basis and in many instances this is represented in the existing Science Order. Hence science teachers teach them currently and will continue to do so. Some examples follow that illustrate how, as a biology teacher, you can contribute to health education. The treatment is illustrative rather than exhaustive. This is preceded by considering the present position on sex education.

Sex education

This is singled out for separate discussion here because it has been dealt with separately in legislation. For maintained schools in England and Wales, the Education (No. 2) Act, 1986 required governors to decide whether sex education should form part of the secular curriculum and to have a written policy statement to that effect. Where sex education was taught, it was required to be done in a way that stressed moral considerations and the value of family life. The 1991 Science Order, teaching which is a binding legal requirement on schools, includes a number of different aspects of sex education.

Human reproduction is mentioned in the PoS at KS1 where pupils find out about the stages of human development. This is taken further at KS3:

> They should study the human life cycle . . . and understand the need to have a responsible attitude to sexual behaviour. (PoS)

At KS4 knowledge of organ systems is extended and pupils also study sex determination.

HIV and AIDS is specifically mentioned in the PoS for Sc2.

Contraception is not mentioned as such in the Science Order, but it might be taught in the context of the section quoted above on the human life cycle and with reference to HIV/AIDS. It could well appear in work involving the study 'of the effects of human activity . . . on the Earth's surface' and 'factors affecting the size of populations of organisms' (PoS for Sc2 at KS3). At KS4 they are required to deal with 'factors associated with a healthy lifestyle and examples of technologies used to promote, improve and sustain the quality of life' (PoS for Sc2).

From September 1994 the position relating to the teaching of sex education is determined by the Education Act, 1993. This requires that:

1. education about human sexual behaviour (including all reference to HIV and AIDS) will be removed from National Curriculum Science;
2. all maintained secondary schools are required to provide sex education including information about AIDS;
3. guidelines about the teaching of sex education are prepared by the governing body of the school;
4. parents have the right to remove their children from the sex education curriculum.

Sex education is discretionary in maintained primary schools. Hence your role as a teacher of National Curriculum Science is restricted to teaching the biology of reproduction. Currently, a draft circular on the teaching of sex education in schools has been produced by the DfE (December 1993) and a policy statement appears in 1994. The details of this statement and the requirements of the 1993 Act will alter the content of the New Science Order, which comes into effect in 1995. Hence the position may not be clarified for some time. In the interim period the best advice to offer is that your teaching of these areas within the science department or outside it is discussed fully with your experienced colleagues. With a sigh of relief you can leave the worrying to them!.

The other eight components of *Curriculum Guidance 5* are, thankfully, much less contentious. However, until we see the new Science Order and start to teach to it in 1995, we cannot be sure about the match between its biology areas and the other eight components of health education. Nevertheless a number of areas of biology will be retained that provide obvious contexts for linking with knowledge and understanding aspects of health education. These are summarized in Table 9.3.

Table 9.3 **Some examples of biology topics that could be used to link with health education**

Health education component	*Main areas*	*Relevant biology topic*
Substance use and misuse	Tobacco, alcohol, other drugs, solvents	Respiratory system, circulation, nervous system (behaviour, reaction times, biorhythm), brewing, fermentation
Family life education	Role of the family, child care, parenting relationships	Reproduction, human development, birth, new techniques (e.g. IVF), puberty, parental care, disease prevention, vaccination and immunization
Safety	Dangers in the environment, keeping personal safety and that of others	Safety in the laboratory, drugs, solvents, allergy, humans in difficulties (first aid)
Health-related exercise	Role of exercise for good health	Physiology of exercise, breathing, heart and pulse rates, fitness
Food and nutrition	Diet and health, food quality and safety	Dietary requirements, role of food constituents, diet and disease, food additives, food production and preservation, monitoring bacterial content of food
Personal hygiene	Cleanliness, reducing disease transmission, body maintenance	Disease causes and their transmission, teeth and dentition, skin and hair
Environmental aspects	Social, psychological economic factors in health; risks	Disease transmission via air and water, waste disposal and recycling, pollution, allergy (e.g. oilseed rape, asthma)

Curriculum Guidance 5 makes some very important points about the teaching methods to be utilized in health education and these offer sound advice for the biology teacher who is looking to extend knowledge and understanding by encouraging individual responsibility, awareness and informed decision-making, which are also aims of health education. Provision of information or attempts to shock youngsters are not, on their own, effective in developing long-term positive attitudes and behaviours. Full participation of pupils is needed via a balanced range of teaching methods. Pupils should:

> assess evidence, make decisions, negotiate, listen, make and deal with relationships, solve problems, and work independently . . . Teaching methods include games, simulations, case studies, role plays, problem-solving exercises etc.[22]

Since many of the issues that are involved are personal and, therefore, potentially sensitive, it is essential that they are dealt with sensitively by the teacher and the group of pupils concerned.

Environmental education

The aims of environmental education have been identified by the NCC in *Curriculum Guidance 7 Environmental Education* (1990) as follows:

> provide opportunities to acquire knowledge, values, attitudes, commitments and skills needed to protect and improve the environment
> encourage pupils to examine and interpret the environment from a variety of perspectives – physical, geographical, biological, sociological, economic, political, technological, historical, aesthetic, ethical and spiritual
> arouse pupils' awareness and curiosity about the environment and encourage active participation in resolving environmental problems.

The Curriculum Council for Wales, in their *Advisory Paper 17 Environmental Education (1992)* describe the purposes of environmental education in this way:

> Ultimately the aim of environmental education is to enable young people to develop a practical capability – an ability to make informed decisions and thoughtful choices which will contribute to the well-being of environment and people, locally and globally. As a result of their environmental learning pupils should be able to clarify their own values and to develop their own personal responses to a range of environmental issues. As well as having developed concerned awareness, pupils should know what action they can take to protect the environment and to promote sustainable use of the Earth's resources – both in their own style of living and through their participation in collective decision-making processes.

The statutory subjects of the National Curriculum, particularly science and geography, but also history and technology, provide the basis for developing environmental education from the details of their respective programmes of study and attainment targets. However it cannot be left to chance that the aims of environmental education will be achieved simply by following the programmes of study of other subjects. A process of planning is needed to provide a framework within which the aims can be addressed.

Curriculum Guidance 7 identified three important components in such a framework:

1. education *about* the environment (knowledge and understanding of key ideas);
2. education *for* the environment (values, attitudes towards, and the capability to take informed action);
3. education *in* or *through* the environment (using the environment as a resource to develop ideas and values).

Education about the environment

Curriculum Guidance 7 lists the knowledge and understanding objectives of environmental education. In the *Curriculum Council for Wales, Advisory Paper 17* these are expanded under three sets of interconnected ideas.

1. Systems and interdependencies. The world is a complex network of interconnected and interdependent systems (ecological, physio-chemical, geological, etc.). Human societies, institutions, systems and cultures are integral components of these global systems.
2. Change and development. Environmental change through time occurs partly by natural processes and increasingly as a result of processes induced by human development – social, economic, and technological.
3. Sustainability and stewardship. Humans are necessarily consumers of the Earth's resources but they carry a major responsibility for maintaining the well-being of the planet and all its inhabitants, human and non-human.

Clearly the science teacher has a considerable contribution to make in teaching these ideas. It is possible to cross-reference these concepts and ideas to particular science statements of attainment. Examples of such cross-referencing have been published for previous and present versions of the Science Order; with another version about to be produced it is pointless to do so here.

Education for the environment

The concerns here are with: ensuring caring use of the environment; finding solutions to environmental problems (acknowledging that there may be different views held by conflicting interests); developing an action capability.

The Curriculum Council for Wales analyses these under the collective heading 'Processes fundamental to environmental learning' shown below:

1. *Enquiry and critical reflection.* This involves, for example:
   (a) collecting, analysing and interpreting information;
   (b) asking questions and defining problems;
   (c) identifying factors affecting the environment;
   (d) appraising evidence and drawing inferences;
   (e) evaluating solutions to environmental problems;
   (f) reflecting critically on the meanings and usage of terms such as 'environment', 'development', 'sustainability', 'stewardship' and 'conservation';
   (g) clarifying and reflecting on one's own values and beliefs and those of others.
2. *Communication.* This involves, for example:
   (a) expressing one's own and others' feelings, opinions, ideas and beliefs verbally (through speech and writing), visually (through painting, drama, dance, etc.), in music;
   (b) listening, observing and describing.
3. *Participation and action.* This involves, for example:
   (a) taking an active part in debate and discussion of environmental issues;
   (b) making personal choices or decisions leading to positive action in response to perceived environmental needs;
   (c) cooperating with others in group tasks (practical problem-solving, decision-making, environmental conservation work).

The match between the section 'Enquiry and critical reflection' and the processes constituting Sc1 is striking. We have mentioned how Sc1 is addressed through specific contexts that are described by the programmes of study for the other attainment targets. The model here is identical: the knowledge and understanding objectives are realized through the process-based approach (methods of enquiry, etc.) summarized above.

Education in or through the environment

If the first two components described above involve the questions 'what?' and 'how?', the third addresses the question 'where?', i.e.

'where should environmental education take place?' *Curriculum Guidance 7* makes this clear:

> First hand experience is an essential part of helping pupils to develop a personal response to the environment and to gain an awareness of environmental issues. This can start in the school itself, in its grounds and immediate locality, progressing to visits to more distant, contrasting localities in the United Kingdom, and in other countries. Those directly involved in environmental issues, such as planners, architects, economists, industrialists, surveyors and politicians, should be encouraged to assist.

The programmes of study for Sc2 provide obvious starting points for basing this work in the school grounds and in local habitats. Some of the other suggestions in the quotation above are rather more challenging and clearly point to the need for a whole-school approach to be taken to environmental education.

Planning, managing and implementing a whole-school approach

Hardie and Hale[24] point out that the links between environmental education and a number of subject disciplines are both its strength and weakness. For pupils' learning to have any coherence a whole-school policy must be developed. In any situation where several subjects are involved, coordination and overview are needed by a senior manager. Guidance about such a policy is available in the publications mentioned. As a class teacher, you could be required to work in one or more of the following ways:

1. alone, contributing in your subject teaching to a coordinated programme involving several specialists in different subjects;
2. in collaboration with teachers from other departments teaching topics, themes, or tasks that are multidisciplinary or interdisciplinary;
3. alone or as a member of a team in a course of PSE.

Some case studies that illustrate these approaches are described in *Curriculum Guidance 7*. Further advice for school-based INSET is available from the Council for Environmental Education.[24]

Notes and References

1. Dearing, R. (1993) *The National Curriculum and its Assessment, Final Report*. London: SCAA.
2. NCC (1990) *Curriculum Guidance 3: The Whole Curriculum.*
3. Dennick, R. (1992) Analysing multicultural and anti-racist science education. *School Science Review*, vol. 72, no. 264, 79–88.

4. Chamberlain, P. J. (1987) Science teaching in multicultural Britain and the SSCR, in Ditchfield, C. (ed.) *Better Science: Working for a Multicultural Society, Curriculum Guide 7*. London: Heinemann Educational Books.
5. Report of the Swann Committee (1985) *Education for All*. London: HMSO
6. See Ditchfield, C. (ed.) op cit.
7. For examples see Vance, M. (1984) *Assessment in a Multicultural Society Biology at 16+*. Longman for the Schools Council; Reiss, M. J. (1993) *Science Education for a Pluralist Society*. Buckingham: Open University Press.
8. Peacock, A. (1990) Building a multicultural dimension into primary science. *Primary Science Review*, vol. 12, 11–12.
9. See Vance, M. (1984) op .cit.
10. Gould, S. J. (1987) Human equality is a contingent fact of history, in *The Flamingo's Smile*. London: Penguin.
11. ASE (1993) *Race, Equality and Science Teaching*. Hatfield: ASE. This is a handbook that includes suggestions for classroom work.
12. For example, Stradling, R., Noctor, M. and Baines, B. (1984) *Teaching Controversial Issues*, London: Edward Arnold; Reiss, M. J. (1993) op. cit.; Wellington, J. J. (ed.) (1986) *Controversial Issues in the Curriculum*. Oxford: Blackwell.
13. Ditchfield, C. (ed.) op. cit.
14. NCC (1990) *Curriculum Guidance 4 Education for Economic and Industrial Understanding*, and NCC (1992) *Science and Economic and Industrial Understanding at Key Stages 3 and 4*.
15. Royal Society (1980) *Biotechnology: A Report of a Joint Working Party*. London: HMSO.
16. From Henderson, J. and Knutton, S. (1990) *Biotechnology in Schools A Handbook for Teachers*. Milton Keynes: Open University Press.
17. After Henderson, J. and Knutton, S. op. cit.
18. Henderson, J. (1987) *Teaching Strategies in Biotechnology, Report of the TRIST 13 to 18 Project*. Sheffield: Division of Education Publication 11.
19. Ingleby, D., Winspear, L., Phipps, R. and Lally, V. (1987) *Biotech*. Hatfield: ASE.
20. See Ingleby, D. *et al*. (1987) op. cit.; Henderson, J. and Knutton, S. (1990) op. cit.; Dunkerton, J. and Lock, R. (1989) *Biotechnology: a Resource Book for Teachers*. Hatfield: ASE; SSCR (1989) *Better Science: A Directory of Resources*. London: Heinemann Educational Books.
21. Teacher Placement Service, Understanding British Industry, Sun Alliance House, New Inn Hall St, Oxford OX1 2QE.
22. A large range of teaching materials are available from the Health Education Council, now the Health Education Authority; see also Dixon, A. (1993) Health education, in Hull, R. (ed.) *ASE, Secondary Science Teachers Handbook*. Hemel Hempstead: Simon and Schuster.
23. Hardie, J. and Hale, M. (1993) Environmental education in Hull, R. (ed.) op. cit. This includes a useful bibliography.

24. Council for Environmental Education (1992) *INSET 5–16 Environmental Education for Science*. Harlow: Longman.

Index